Natural Selection: Domains, Levels, and Challenges

GEORGE C. WILLIAMS

Department of Ecology and Evolution
State University of New York at Stony Brook

D0104166

New York   Oxford
OXFORD UNIVERSITY PRESS
1992

Oxford University Press

Oxford New York Toronto
Delhi Bombay Calcutta Madras Karachi
Petaling Jaya Singapore Hong Kong Tokyo
Nairobi Dar es Salaam Cape Town
Melbourne Auckland

and associated companies in
Berlin Ibadan

Copyright © 1992 by Oxford University Press, Inc.

Published by Oxford University Press, Inc.,
200 Madison Avenue, New York, New York 10016

Oxford is a registered trademark of Oxford University Press

Library of Congress Cataloging-in-Publication Data

Williams, George, C. (George Christopher), 1926–
Natural selection : domains, levels, and challenges / George C.
Williams.
p.   cm. — (Oxford series in ecology and evolution ; 4)
Includes bibliographical references and index.
ISBN 0-19-506932-3. — ISBN 0-19-506933-1 (pbk.)
1. Natural selection. I. Title. II. Series.
QH375.W52   1992
577.01'62—dc20
91–38938

9 8 7 6 5 4 3 2

Printed in the United States of America
on acid-free paper

for Doris

Preface

I started gathering ideas and references for this book during my 1981–2 sabbatical leave as a Fellow at the Center for Advanced Studies in the Behavioral Sciences at Stanford, California. I began the writing in earnest on my next sabbatical (1988–9) at Queen's University, Kingston, Ontario. That year I was assisted by a generous grant from the J. S. Guggenheim Foundation. I am immensely grateful to all of these institutions, and especially to the State University of New York at Stony Brook for the sabbatical leaves and other assistance without which the work would have been impossible.

I owe a great debt to many individuals. Eric L. Charnov, Mary Jane West Eberhard, and David Sloan Wilson read a preliminary draft of the manuscript and gave generously of their time and many valuable criticisms. Their efforts far exceeded normal expectations for the critical review of a colleague's writings, and the work is much improved as a result. Other valuable suggestions were offered by Series Editors Paul H. Harvey and Robert M. May. Richard A. Watson offered valuable comments on Chapters 1 and 2. David N. Reznick generously provided Fig. 7.2, and Radford Arrindell made it possible for me to photograph a specimen in his charge for Fig. 9.1. Junhyong Kim contrived the graphics routine needed for Fig. 9.2. My wife, Doris Calhoun Williams, contributed her bibliographic wizardry and helped in innumerable other ways. I am grateful to Cornell University Press and translator Margaret Talmadge May for permission to reprint part of Galen's "On the Usefulness of the Parts of the Body".

New York   G.C.W.
October 1991

Contents

1 A Philosophical Position, 3

2 The Gene as a Unit of Selection, 10

2.1 A general model of selection in the codical domain, 13
2.2 Selection at the level of the gene, 16
2.3 Kin selection, 19
2.4 Effects of inbreeding and asexual reproduction, 21

3 Clade Selection and Macroevolution, 23

3.1 Requirements for selection among gene pools and clades, 24
3.2 Selection of genes vs. selection of gene pools, 27
3.3 How important is clade selection?, 31
3.4 Characters likely to be important in clade selection, 34

4 Levels of Selection Among Interactors, 38

4.1 Adaptation as the material effect of response to selection, 38
4.2 Natural selection within organisms, 41
4.3 Selection of individuals in populations, 43
4.4 Trait-group selection, 45
4.5 Selection of populations within species, 46
4.6 An aside on beehives and haystacks, 48
4.7 Selection among more inclusive phylads, 50
4.8 Clade selection and punctuated equilibria, 53
4.9 Genealogically mixed interactors, 54

5 Optimization and Related Concepts, 56

5.1 Frequency-dependent selection, 56
5.2 Parameter optimization, 60
5.3 Character values and fitness values, 65
5.4 Strategies, tactics, and winnings, 69

6 Historicity and Constraint, 72

6.1 The organism as historical document, 72
6.2 Natural selection and phylogenetic constraint, 76
6.3 Developmental constraints, 80
6.4 Genetic constraints, 85
6.5 Unity of type and *Bauplan*, 87

7 Diversity Within and Among Populations, 89

7.1 Natural history as the foundation of comparative biology, 89
7.2 Causes of variation among individuals, 91
7.3 Examples of variation within a population, 94
7.4 Cladogenesis, 98
7.5 Comparative biology without functionalism, 100
7.6 Functional interpretations of phylogenetic divergence, 101

8 Some Recent Issues, 106

8.1 The lek paradox, 106
8.2 The female pheromone fallacy, 111
8.3 *Schreckstoff*, 113
8.4 The helpful-stress effect, 116
8.5 Species fallacies, 118

9 Stasis, 127

9.1 Taxonomic stasis, 128
9.2 A desperation hypothesis, 132
9.3 Character stasis, 135
9.4 Avian and mammalian body temperature, 136
9.5 Electrolyte concentrations of marine vertebrates, 138
9.6 Why no viviparous birds or turtles?, 139
9.7 Other problems of character stasis, 141

10 Other Challenges and Anomalies, 143

10.1 Haldane's dilemma, 143
10.2 Paradoxes of sexuality, 148
10.3 Other difficulties, 150

References, 154
Appendix (excerpts for Galen and Paley), 190
Index, 203

Natural Selection

1

A philosophical position

Successful biological research in this century has had three doctrinal bases: *mechanism* (as opposed to vitalism), *natural selection* (trial and error, as opposed to rational plan), and *historicity*. This last is the recognition of the role of historical contingency in determining properties of the Earth's biota. The formation of the Earth was a unique event, and unique events have been altering evolutionary processes ever since, always keeping futures unpredictable from presents. As S. J. Gould (1989, p. 48) forcefully expressed it, if we could rewind the tape of evolutionary history to the remote past and play it again, it would turn out entirely different. The term *evolution* in its original sense of an unfolding or development, analogous to the development of an individual animal, is misleading (Salthe 1989).

Mechanism implies that only physico-chemical processes are at work in an organism. Every vital function is performed by material machinery that can in principle be understood from a physical and chemical examination. The opposed doctrine of vitalism maintains that the observable machinery has but limited autonomy and is controlled by a purposive entity peculiar to living organisms. The history of much of biology in the nineteenth and early twentieth centuries can be viewed as a gradual retreat of vitalism and (final?) triumph of mechanism. Authors of biology textbooks that I read in the 1940s often presented vitalism as an idea with some adherents, but implied that mechanism was nearly universal among practicing professionals.

Vitalism was notably persistent in experimental embryology. Driesch (1929) and other influential researchers thought that the kind of goal-directed control shown by development could not arise from the observable structures themselves. The controlling principle must be found in vital forces (Driesch's *entelechies*) found only in living organisms. Even well-trained scientists in Driesch's time had but limited experience with negative feedback control loops and no way of imagining today's elaborate cybernetic technology. Modern embryologists must find it easier to imagine that the physical structure of an embryo embodies all its own regulatory mechanisms.

Vitalism today can be recognized in the psycho-physical dualism of some neural and behavioral biologists (e.g. Griffin 1981), who claim that explanations must make use of explicitly mental factors in addition to the merely physical. Griffin's concession (1981, p. 24) that the mental depends on the physical is difficult for me to interpret. If he means total dependence there is no longer any reason to make use of mentalism in biological explanation. If he means less than total, the shortfall becomes the realm of Driesch's (1929, p. 301) 'entelechy, as a natural agent, acting upon the matter of the body.'

An open mind must be kept on the issue of vitalism, especially for neural phenomena. There could be no scientific finding more important than the demonstration of a mental principle, not itself physical, but capable of altering cause–effect relations in neural machinery. Such a demonstration would at one stroke compromise mechanism and make possible a scientific study of the mind–body relation (mind–brain, in the last two centuries) which has justifiably engaged philosophers at least since Descartes. My guess is that this discovery will not be made, and that we will stay stuck with vague ideas of psycho-physical parallelism through time, the one descriptor that the mental and physical have in common (see Maugh 1981; Shallice 1988; Colgan 1989; Corbetta *et al.* 1990, for recent discussions). I know how I feel when I claim 'Cogito ergo sum,' and by intuition and analogy I assume that Descartes felt the same way. I doubt that I will ever have any more direct insight into Descartes' or anyone else's thinking.

I am using such terms as *mind* and *mental* in the usual sense of our own view of our own consciousness. Any sane individual intuitively attributes such a consciousness to, at the very least, other sane adults and children of verbal age. This intuitive concept of mind implies nothing about capabilities for information processing. Even the cheapest pocket calculator is much better than I am at processing certain kinds of information. Most people do not, in serious discussion, attribute consciousness to a calculator or even a mainframe computer. I have no inclination to deny the mental realm or belittle its philosophical importance. I am inclined merely to delete it from biological explanation, because it is an entirely private phenomenon, and biology must deal with the publicly demonstrable. It is worth noting that Descartes did not say 'Cogitamus ergo sumus.'

A fantasy version of the Turing test (Langton 1989; Denning 1990) may help to show the irrelevance of mentalism to biology. An utterly strange object, clearly an extraterrestrial visitor, lands on the village green. With arm-like appendages it starts sampling its environs, picking up grass, dead leaves, beverage containers. Local authorities seize it and take it to the laboratory for analysis. Among the questions to be posed is that of the object's guidance system. Possible answers are: (1) it is or contains a robot designed especially for this mission; (2) it is or contains a general-purpose robot; (3) it is or is manned by a real organism, one that has evolved advanced capabilities for processing information, but no consciousness like that which we perceive in ourselves; (4) it is or is manned by an organism with intelligence and consciousness closely analogous to our own.

In principle it should be possible to determine whether our visitor has capabilities beyond those needed for a single task. If so, the first possibility is ruled out. An organism evolved by natural selection can be distinguished from an efficient multi-purpose robot by its having, not only its manifest capabilities attributable to natural selection or purposive engineering, but also many arbitrary or malfunctional features attributable to historical legacy (discussed mainly in Chapter 6). I would claim that there is no set of tests that could possibly decide between the third and fourth possibilities (detailed arguments in Churchland 1986). If I am right, there is no way in which recognition of consciousness in our visitor, or in an experimental organism, could help us to understand it objectively and scientifically. Such help would be forthcoming only if, as suggested above, consciousness can be shown to be an immaterial principle capable of altering cause–effect relations in material systems.

The second basis of modern biology is the assumption that the Darwinian process of natural selection accounts for all aspects of the adaptation of an organism to a particular way of life in a particular environment. Natural selection is a system of corrective feedback that favors those individuals that most closely approximate some best available organization for their ecological niche. For Darwin this was a way of explaining the great diversity of organisms that inhabit the Earth. Biologists today use it more often to explain the great generation-to-generation stability of a species' characteristics, or the near absence of evolution. They assume that an organism is already close to some maximum achievable level of adaptation, and on this basis they attempt to predict features not yet known. This research strategy, the *adaptationist program* of Gould and Lewontin (1979; see also Mayr 1983; Williams 1985; Horan 1989), has been in use for centuries. It has been increasingly common and more explicitly based on natural selection during the last few decades.

In the usual textbook accounts, and in most applications by professional biologists, natural selection operates by the reproductive successes and failures of organisms in populations. Shifting gene frequencies keep the record of success and failure because greater reproductive success for any individual means greater prevalence of its genes in the future. There is nothing in this process that can anticipate future needs or foster adaptations for the good of the population or species or other collective entity. As I will argue in Chapters 3, 4, and 9, it is logically possible for selection to operate at group levels to produce adaptive group organization, and I suggest that certain sorts of group selection are probably important.

In practice, higher levels of selection are seldom invoked, and biologists routinely predict and find that the properties of organisms are those expected if selection operates mainly on the varying capabilities of individuals. This perspective has prevailed only in the last 30 years. Prior to the 1960s, even explicitly evolutionary discussion was often vague and uncritical in its use of the adaptation concept, and benefit-to-the species modes of thought were common. Routine use of natural selection as a hypothetico-deductive basis for understanding adaptation is rather new.

According to Maynard Smith (1988*b*, Chapter 28) and Ruse (1989), mechanistic biologists assume an *organism-as-crystal* and adaptationists an *organism-as-artifact* concept. An *organism-as-document* approach should also be recognized for biologists interested mainly in unique evolutionary histories. The doctrine of historicity holds that the properties of the organic world, from the total biota to minute parts of animals and plants, are the results of unique historical contingencies, some of which are documented in features of the organisms themselves. Historicity contrasts with any developmental view of the diversification of life on Earth. In a developmental process, a moderate disturbance is countered by a control mechanism that prevents the disturbance from redirecting the process. While individual morphogenesis is eminently developmental, organic evolution, in current orthodoxy, is entirely historical. Every disturbance redirects it to at least a minor degree initially, and with continually ramifying effects thereafter. A major disturbance is likely to have major effects immediately and on all subsequent history.

An example of a major event is the appearance, a few million years ago, of the Isthmus of Panama, which joined two formerly independent terrestrial biotas and divided a formerly continuous marine one. In the resulting Great American Interchange (Marshall *et al.* 1982; Vermeij 1987) many terrestrial organisms extended their ranges, gave rise to new species, or went extinct from competition or predation. Minor events such as a particular mutation in a particular individual at a particular time could have important effects on all descendants of the mutant individual and all others that interact with these descendants. An analogy is the *butterfly effect* in meteorology: the flapping of one butterfly's wing now will make the weather a month from now different, everywhere on earth, from how it would have been without the flapping (Gleik 1987, Chapter 1).

Many features of living organisms are functionally arbitrary or even maladaptive. The neck skeletons of giraffe, man, and mouse are all marvels of mechanical engineering for the different ways of life of these divergent mammals. Yet all have seven vertebrae in this region, a functionally inexplicable uniformity. The only acceptable explanation is historical, descent from a common ancestor with seven cervical vertebrae. That ancestor must have reached a threshold of specialization (e.g. of cervical nerve contributions to the brachial plexus) that would make any mutant individual with six or eight cervical vertebrae functionally defective in serious ways. Any that might have been formed were consistently eliminated by natural selection.

The same necks can illustrate persistent maladaptation. All vertebrates are capable of choking on food, because digestive and respiratory systems cross in the throat. This likewise is understandable as historical legacy, descent from an ancestor in which the anterior part of the alimentary tract was modified to form a previously unneeded respiratory system. This evolutionary short-sightedness has never been correctable. There has never been an initial step, towards uncrossing these systems, that could be favored by selection. Other examples, and a more detailed treatment of historicity, are given in Chapter 6.

Mechanism is generally understood by biologists and to some extent by educated laymen. There is little grasp of natural selection or historicity outside biology. Within biology they are well understood only by those especially interested in evolution. The prevalent ignorance of these two facets of Darwinism is clearly shown by what is considered plausible by readers and writers of science fiction. Nothing unreasonable is seen in having functional improvements such as increased strength or new sensory capabilities result from a single mutation. Space travelers routinely find copies of a unique historical document, *Homo sapiens*, on more than one planet in the universe. Perhaps science fiction is beginning to rise above such absurdities. Gould (1989, pp. 285–8) identified some examples of appreciation of historicity in recent popular fantasies, both literary and cinematic.

This book is written from a belief in the continuing validity of the three conceptual foundations of biology, and from a conviction that these three are sufficient at the most basic level. Other useful principles are secondary and of limited scope. The crossing of digestive and respiratory systems is universal only in the vertebrates. Mendel's laws have operated only since the preCambrian origin of the necessary chromosomal machinery, and only in the descendants of those lineages in which this origin took place. I assume that the same three principles, and no others, will be found to apply everywhere in the universe where life has arisen. This is Dawkins' (1983) idea of *universal Darwinism*. Of course any of my three basic principles might be wrong. I suggested as much about mechanism when I discussed the possibility of an immaterial but physically active principle of mentality.

Might there be somewhere a planet on which the biota arises and becomes more complex deterministically rather than historically? The Earth's biota may have developed in a nearly predetermined fashion prior to the origin of mechanisms for maintaining physical boundaries to individuals. The Pliocene mixture of North and South American mammals across the Isthmus of Panama produced interesting results only because the northern and southern faunas contained recognizably distinct groups of mammals. Permanently distinct species or other taxonomic groups can arise only if there are mechanisms for keeping them separate. In a primitive amorphous biosphere, an adaptive change in one place might become the property of the whole system, as long as that biosphere erects no barriers to such spread. The origin of the barriers is the origin of individuals and of the clones that result from their division.

I am supposing the existence of a planet that supports life but no organisms—vital molecular processes arose but individuality did not. Even here, chance events must sometimes have conditioned all future change. At a given moment, there may have been two possible mutations in self-replicating molecules, either of which would have been favored by selection, but not both. Whichever chanced to occur and spread first would alter conditions so that the second could no longer be favored. Purely by chance, the biota would have taken one evolutionary path rather than another. I predict that no extraterrestrial biota will ever be found in which the universal principles of mechanism, natural selection, and historicity will not prevail.

There are many other philosophical issues on which a biologist must necessarily have a position. Among these are concepts of causality and the nature of explanation. Mayr (1976, Chapters 23 and 26, 1982, Chapters 2 and 3) gives a thorough discussion of these matters from a biologist's perspective. Sherman (1988, 1989) has made an important advance on the traditional distinction between proximate and ultimate causes in biology. My favorite discussion of biological cause and effect is that of Glass (1963, reprinted 1985, pp. 51–6). He points out that processes found in intermediary metabolism show several important kinds of cause–effect relationships unknown or unnoticed by physical scientists, and they show the fallacy of some classical cause–effect definitions. Biologists (like engineers) are endlessly confronted by historical causation and by positive and negative feedback loops not normally considered by physicists, or by philosophers for whom physics is the exemplar of science. Related to the cause–effect issue, and also to natural selection and historicity, is that of progress in evolution, a concept for which I have little use (Williams 1966*a*, Chapter 2) for reasons discussed recently by Provine (1989) and Gould (1989).

There are also questions of the ethical implications of evolutionary ideas and the issue of reductionism vs. holism, on both of which I have already had much to say (Williams 1985, 1988, 1989, 1992). Here I will merely add an endorsement to D. S. Wilson's (1988) complaint that the terms *holism* and *reductionism* have been used in quite different senses by different authors. In most discussions I have heard, the distinction depends mainly on the sizes of glassware one uses in research. Those whose laboratories are full of specimen jars and aquariums are holists, those with test tubes and petri dishes reductionists. I work more with jars and aquariums but prefer to ally myself with the reductionist camp, mainly because holism is too often a seemingly respectable label for mysticism and pedantic obfuscation. I discuss holism and reductionism more seriously in Chapter 4.

2

The gene as a unit of selection

Well-reasoned treatments of Darwinism for wide audiences (e.g. Sober 1984; Lloyd 1988; Brandon 1990) have recently had the benefit of Dawkins' (1976, 1982*b*) and Hull's (1988) concepts of *replicators* and *interactors*. The concepts are general in scope, but replicators usually mean genes, and interactors the individual organisms that transmit their genes in varying degrees to future generations. Organisms do the reproducing; genes direct development and provide the heritability on which response to selection depends.

I believe these discussions to be a great improvement over many that went before but that the replicator–interactor distinction is often used in ways that subvert much of its potential usefulness. As a remedy I suggest that these terms should refer to two mutually exclusive *domains* of selection, one that deals with material entities and another that deals with information and might be termed the *codical* domain. A unit of selection in the codical domain would be a *codex* (*codices* is the only plural in my dictionary).

The material and the codical are separate domains because of a dearth of shared descriptors. In discussing the codical domain we use such terms as bits, redundancy, fidelity, and meaning. The material domain is described by color, charge, density, volume, etc. The only descriptor they have in common is time, and events in one domain can be established as before, after, or simultaneous with events in the other. A catastrophic act of arson in the seventh century was a physical event at exactly the same time as the destruction of information in the library at Alexandria. Information can exist only as a material pattern, but the same information can be recorded by a variety of patterns in many different kinds of material. A message is always coded in some medium, but the medium is really not the message.

The usefulness of maintaining a conceptual separation of these domains is clearest for cultural evolution. A century ago people followed the adventures of an 'Ingenioso hidalgo' named *Don Quixote*, loudly urged *Cherubino alla vittoria* and decorated conifers every December. They still do these same things because they are still expressing the same cultural replicators. *Don Quixote* is information, most often coded as a pattern of ink on paper, but it can exist in many other media. It is often transmitted visually and stored ephemerally in human brains. It has no doubt been recorded on disks and magnetic tapes for transduction into sound for transmission to brains of people unable to read ink on paper. In any medium, *Don Quixote* can form an archive from which copies can be made to any other medium, but no matter what the medium, it is always the same book.

It has not been customary to think of genes as analogous to books in this respect, but I think it helpful to do so, despite the more serious constraints on gene replication. It is indeed a fact of life that information in proteins and RNA is not routinely transcribable into the medium of DNA. Only DNA provides the durable archive for most of the Earth's organisms. This constraint should not blind us to the fact that it is information we are concerned with, and that DNA is the medium, not the message. A gene is not a DNA molecule; it is the transcribable information coded by the molecule.

The idea that the gene is a package of information, not an object, is neither new nor idiosyncratic. It is explicitly argued by Mosterin (1986, 1988), by Gouyon and Gliddon (1988), and by Williams (1985). It is implied by many earlier works as in Dawkins' (1976, pp. 29–30) analogy between a gene and a page in a book, Dretske's (1985) general distinction between information and its medium of expression, and Williams' (1966*a*, p. 33) reference to the gene as a *cybernetic* entity. It is supported by the facts of molecular biology. The genetic code resides entirely in the base-pair sequence. The ribose chain is the same in every DNA molecule, just as the same glue can be used in the binding of quite different books. Tertiary structure dependent on pH or other conditions, and any alterations reversible by repair mechanisms, are not transmitted or transcribed.

A DNA molecule is matter, with mass and charge and length, and just as much a part of the phenotype as a molecule of visual pigment that it may play a role in synthesizing. Nucleotides and their component atoms can come and go freely, but as long as each replacement is by the same chemical entity, the information, and therefore the gene, remain the same. It should also be borne in mind that a whole active organism shows a constant material flux with its environment. Its continuity is of pattern, not substance (Johnson 1987). It is not really an object, it is a region where certain processes take place. Only the fact that some processes may seem slow, like the shedding of epidermis or the replacement of calcium atoms in bones, makes an organism-as-object concept useful in some contexts.

While I think it important to keep separate the codical and material domains, I would not insist that the codex concept completely replace the idea of a replicator. Perhaps the term *replicator* could apply only when the same message is copied into the same medium. One DNA molecule making another just like itself would be replication. So would a one-sheet photocopy used to make another photocopy in the same machine. Photocopying a book page would perhaps not be replication in this sense, because the copying machine would use a different sort of paper and ink. This kind of photocopying might be considered transcription to a different medium. This is a topic that warrants serious thought.

Brandon (1988, 1990) and Sober (1984) concede the usefulness of the replicator–interactor distinction but consistently regard replicators as material objects and miss the codex concept. Thus they regard not only genes but whole chromosomes and even clonable protozoa as both replicators and interactors. Interactors they surely are, and if two daughter chromosomes really do have exactly the same chemical composition as the original parental chromosome, despite environmental changes that might have occurred, chromosomes may also be replicators in my special sense. Surely this would not be true of whole organisms reproducing asexually. No daughter would be exactly the same materially as the mother. It is more useful to think of all these examples only as interactors, specified by and transmitting the message coded by base-pair sequences. Only this message is the codex, and it is the same message whether written in DNA, RNA, or protein.

Or beyond protein. Transcription of genetic information means imposing a pattern on a material medium, and there is no theoretical limit to such chains of transcription. Dogs may be able to distinguish closely related human associates by smell, but not ordinarily monozygotic twins. Bonner (1980, p. 111) confessed to being 'tempted to say that the dog seems to be able to smell our genes.' I suggest that this statement may be taken a bit more literally than Bonner probably intended. That human genetic differences can cause differences in the behavior of dogs is an example of what Dawkins (1982a) meant by 'extended phenotype.'

For genetic information the usual practice is to distinguish transcription in the development of an individual from replication in its sexual reproduction. This distinction is often less applicable with cultural information or for vegetative reproduction. Transcription and replication are examples of the general process of *proliferation* of codices, for which there is no theoretical limit. There is no limit to the number of dogs that can have their behavior influenced by a human genotype or the number of future individuals that can bear the genes of that genotype. This lack of a conservation principle for information, analogous to the conservation of matter or energy, was called the *xerox principle* and attributed to Dretske (1981) by Ward (1989). I prefer *Dretske principle* because I think it more appropriate to name an important idea for its originator than for a trade-name used in explaining it.

For natural selection to occur and be a factor in evolution, replicators must manifest themselves in interactors, the concrete realities that confront a biologist. The truth and usefulness of biological theory must be evaluated on the basis of its success in explaining and predicting material phenomena. It is equally true that replicators (codices) are a concept of great interest and usefulness and must be considered with great care for any formal theory of evolution, either cultural or biological. Van Valen (1982) argued persuasively for a primary focus on the idea of continuity of information as a basis for the biological concept of homology.

Continuity of information is also a fundamental idea for the study of culture. There is more intellectual interest in the survival of *Don Quixote* as information than in survival of particular ink patterns in paper books or of magnetic orientations on tapes. Holders of copyrights are more covetous of their books as information than as objects. They demand control of replication in any medium, not just ink on paper. Someday perhaps ink on paper will be obsolete as a medium in which to express books. People may do all their reading on liquid-crystal monitors. Some twentieth century bibliophiles may deplore such a development, but surely they would prefer it to the loss of the information content of *Don Quixote* and other books now in our libraries.

2.1 A general model of selection in the codical domain

Information can proliferate and be edited by natural selection only if the selection affects the information at a greater rate than competing processes such as mutation and drift. A given package of information (codex) must proliferate faster than it changes, so as to produce a genealogy recognizable by some diagnostic effects. A good test of the susceptibility of an entity to natural selection is whether its history can be modeled realistically by a dendrogram (Fig. 2.1). The branching lines in the figure could represent the history of a recognizable *meme* (Dawkins' (1976) cultural analog of the gene), such as the Christmas tree idea. A branching could represent the successful spread of the idea from one family to another, or from parents to offspring who later continue the practice in their own households.

The same branching diagram can also represent successful passage of a gene from parent to offspring. Its transmission can be traced because

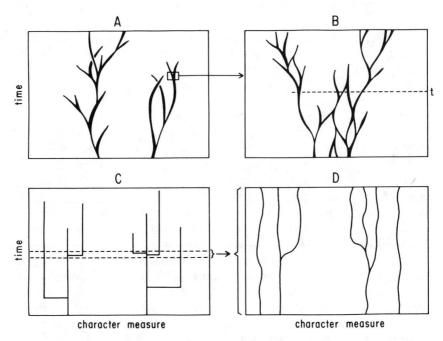

Fig. 2.1. Dendrograms to illustrate the natural selection of codices (units of information). A is the usual sort of phylogenetic representation, perhaps of species in two related genera, with line thickness indicating numerical abundance. The genus to the left would seem to be favored by clade selection, but the effect would not be significantly different from random branching and extinction.

B is a detailed enlargement of part of A, with the lines representing gene pools of local populations rather than species. The lines not only branch, they sometimes anastomose, as expected of lineages that have not yet attained the rank of separate species. An observer at time *t* could identify five genetically independent lineages, each with its own distinctive history, gene frequencies, and ecological circumstances. There would no way of knowing that one gene pool will shortly divide into two, and two others fuse into one.

The two remaining diagrams show only existence through time, with no indication of abundance. C is the sort of rectangular dendrogram often used to illustrate the pattern of *punctuated equilibria* sometimes recognized in fossil records. D represents the part of C between the dotted lines. Most biologists assume that D would emerge from C with sufficiently precise information on average character values and a sufficiently small-scale time resolution for a fossil record. C can also realistically represent the transmission of genes through time, with each branch point representing a mutation, and the horizontal axis a codical feature such as bytes constrained by some functional consideration. C can equally well represent the persistence of genotypes, and their occasional change from mutation, for organisms with exclusively clonal reproduction.

When dendrograms are used to illustrate selection in the codical domain, continuous lines should be used only for information continuously present. They should not be used for entities, such as genotypic or phenotypic categories in Mendelian populations, that recur rather than persist.

14

it produces phenotypic effects different from those of alternative alleles, just as the Christmas tree meme's effects can be distinguished from those of the competing *bah humbug* meme. In considering various possible kinds of selection later in this and other chapters, I will frequently refer to Fig. 2.1 and invoke the dendrogram test.

Information, unlike objects, can be recognized as having meanings of various kinds, including moral meanings. Machiavelli's *The Prince* is a codex that might be described as a set of directions for taking unfair advantage of subordinates. I might reasonably point to a copy and term it an immoral book. If so, I would be generally understood to be talking about the message, not the medium. I would not be implying that paper and ink and glue can be condemned as immoral. Likewise when Dawkins (1976) refers to a selfish gene, he implies that we condemn a message, not some picograms of DNA, although this is somehow the interpretation given by Midgely (1985). This absurd error must result from her intrusion of mentalism (pp. 4–5) into biology. Surely Dawkins had no wish to imply that a tiny trace of DNA could be an entity that could have a mental experience of malice or guilt. He might well recognize it as the bearer of a selfish message: Exploit your environment, including your friends and relatives, in ways that maximize my proliferation.

Genes and memes differ in mode of transmission: genes exclusively from parent to offspring, memes between any associated individuals. This difference has generally been recognized in treatments of the natural selection of cultural traits (e.g. Bonner 1980; Lumsden and Wilson 1981; Boyd and Richerson 1985), but its importance has not. This is shown by the theorists seeking biological analogies mainly in the realm of population genetics. Indeed, Bonner (1980, p. 18) claims that an inference formally parallel to Fisher's fundamental theorem can be drawn for cultural evolution. I suspect that this is true only for the special case of cultural elements transmitted exclusively to descendants.

Biological analogies for human cultural evolution may be sought more appropriately in the field of epidemiology, a possibility recognized but not adequately discussed by Cavalli-Sforza and Feldman (1981, p. 53). Population genetics, in fact, may be defined as that branch of epidemiology that deals with infectious elements transmitted exclusively from parent to offspring. This makes the long-term fitness of the infectious elements entirely dependent on the fitness of hosts, a standard epidemiological conclusion (Ewald 1988; Bull *et al.* 1991). Fisher's theorem, optimization models (Chapter 5), and related devices depend critically on the exclusively vertical transmission of genetic materials.

There is no more reason to expect a cultural practice transmitted between churchgoers to increase churchgoers' fitness than there is to expect a similarly transmitted flu virus to increase fitness. When rates of transmission of infectious elements are rapid relative to turnover rates in host populations, the fitness of the transmitted information is largely decoupled from that of hosts. Cultural pathogens (Cavalli-Sforza and Feldman 1981, p. 31) are to be expected, as the prevalence of astrology, substance abuse, and other nugatory practices demonstrates.

2.2 Selection at the level of the gene

The discussion in this section relates mainly to the kinds of organisms of primary interest to a majority of biologists: most vertebrates, arthropods, and roundworms, many semelparous and some iteroparous plants. It assumes that there is limited inbreeding and no asexual reproduction. Later I broaden the discussion to deal with clonal or closely inbred lineages.

The codex in most discussions of evolution has been the gene, for good reasons, as many have recognized (e.g. Dawkins 1976, 1982*b*; Sterelny and Kitcher 1988). A serious conceptual complication is the fact that genes are ordinarily associated into genotypes, which collectively direct the development of a unique individual, the usual kind of interactor assumed in discussions of evolution. The genes in a genotype are transmitted to the next generation at a rate (usually zero) determined by environmental effects on the interactor. These complications are best handled by regarding individual selection, not as a level of selection in addition to that of the gene, but as the primary mechanism of selection at the genic level. Because genotypes do not replicate themselves in sexual reproduction (can not be modeled by dendrograms), they can not be units of selection. I believe it important for conceptual clarity to use *individual selection* in this sense only. It is selection within a Mendelian population and is based on the varying fitness of genotypes.

Some essential concepts for a model of selection at the level of the gene are the individual, the gene pool, and the relationship between them (Fig. 2.2). The individual arises as a unique temporary genotype formed in sexual reproduction by a sampling of gametes from the gene pool. The genotype thus formed is a set of instructions for producing an interactor, the fitness of which will be determined mainly by how favorable an environment it encounters during development. For most individuals formed from a gene pool, fitness will soon be zero, because environmental stresses will destroy them before maturity. Of those that mature, fitness may vary widely, and individuals of similar fitness may have greatly dissimilar reproductive success because of chance environmental events. Reproductive success is measured by the magnitude of the interactor's lifetime total of genes put back into the gene pool, relative to the performance of others in the population.

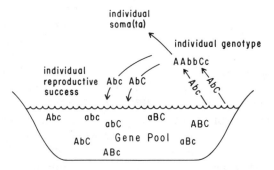

Fig. 2.2. Relationships among the concepts of *gene*, *genotype*, *gene pool*, and individual *soma*. Only the soma represents a concrete object. The other entities exist as information (the codical domain). The differential success of individual genotypes in contributing genes to a common gene pool, as shown here, is the only legitimate meaning for the term *individual selection*.

The measurement of reproductive success is complicated, both pragmatically and conceptually. A mere count of offspring produced will not be an accurate measure of ultimate success, unless one also knows the fitness of those offspring and the average number and fitness of the offspring of competing members of the population. A given level of real success in a growing population demands a higher level of apparent success than it would in a decline. In reality, these conceptual complications are seldom of serious concern. One is not ordinarily interested in the evolutionary success of a particular individual. The measure of interest is the average success of individuals that bear a certain gene, or a certain trait that implies genes different from those in individuals with an alternative trait.

It is also often assumed that increases and decreases in population size can be ignored, on the assumption that population change must always average nearly zero on an evolutionary time scale. Thus the second term in Fisher's (1930, p. 27) expression for reproductive value is frequently ignored (e.g. Williams 1966*b*; Werner 1986; Charnov 1989), and the long-term average (zero) is treated as a constant. I am not aware that anyone has seriously modeled the evolutionary effects of a high variance in Fisher's Malthusian parameter, although there is much relevant discussion in Charlesworth (1980) and Orzak (1985).

The recognition (Fig. 2.2) that the soma is a material entity is not to be confused with a claim that it is really a static object. On a scale of microseconds a candle flame would seem to be an object. On a scale of seconds it is a place where certain processes occur. Likewise a human soma is an object if viewed for seconds or hours, but not over a period of years or decades (Johnson 1987). Like a candle flame, it is a region where substances enter, play various roles in various processes, and later depart, usually in altered chemical forms. The persistence of somatic pattern over years and decades is not a material persistence like that of a robot; it is the persistence of information, partly genetic and partly taken from the environment.

Use of a dendrogram (Fig. 2.1C) to model success and failure in genic selection implies a definition of the term *gene*. A gene is that which reliably survives the process of meiosis intact. An equivalent expression is to say that a gene is that which obeys Mendel's law of independent assortment. *Independent* means *independent of other genes in its transmission* through generations. There is no implication of functional independence. The genes of a genotype may and do interact in complex ways in their direction of developmental processes in an individual. They interact with both their genetic (Waters 1991) and ecological environments in determining fitness.

The time scale in all of the dendrograms in Fig. 2.1 should be considered long relative to events that determine success and failure. A millimeter on a gene dendrogram (Fig. 2.1C) might correspond to a generation in nature. A gene would often pass through many different genotypes (genetic environments) and life histories between its origin and extinction. Its effective environment is multigenerational and fine grained, as noted by Frank and Slatkin (1990) and by Sterelny and Kitcher (1988).

A gene is generally considered to be favorably selected if it increases in frequency faster than expected from random factors such as drift and mutation pressure. The conventional measure of an allele's frequency is the sum of its homozygotes and half its heterozygotes divided by the total population. This is a convenient and satisfactory metric for most purposes, but I think it should be regarded as an approximation to what we should really mean by favorable selection. The selected genes should be increasing their control over matter. Not all homozygotes and heterozygotes should be counted equal. The genes of a large, vigorous, rapidly metabolizing and prolifically reproducing individual are having more of an effect on matter than those of a less influential individual. A rather similar view is implied by Van Valen's (1989) idea of fitness as the maximization of 'expansive energy.'

There are analogous problems in cultural evolution. Two recipes, to borrow an example from Dawkins (1976), may be about equally prevalent in cookbooks and cardfiles yet be quite unequally successful in imposing pattern on matter. One may be new and not have had enough time to appear in a large number of cookbooks, but it may already be in frequent use. Maybe a hundred cakes a day are baked according to its directions. The other is mainly in old, seldom used cookbooks, and days may go by without its resulting in even one cake. Equality in the medium of ink on paper need not imply equality in total patterning of matter. A recipe is a useful example in another respect. It is much easier to use the recipe to produce the cake than to use the cake to produce the recipe. There seem to be inequalities in the ease of transcription between media in the cultural realm analogous to the inequality in ease of transcription of DNA to RNA and RNA to DNA.

2.3 Kin selection

If individuals regularly interact with relatives, selection at the level of the gene will depend in part on effects on the relatives' reproductive success. The resulting conceptual complications led to the theory of *kin selection*, a term introduced by Maynard Smith (1964). The individual property selected for when relatives regularly interact is what Hamilton (1964) called *inclusive fitness* (Fig. 2.3). Kin selection is not a kind of selection in addition to individual selection, nor a special level of selection, certainly not a form of group selection, as is sometimes claimed (E. O. Wilson 1975; Wade and Breden 1981).

As explained in the next chapter, recurrent groupings of relatives can provide a basis for trait-group selection, an idea effectively championed by D. S. Wilson (1980). This would not normally mean that attention to

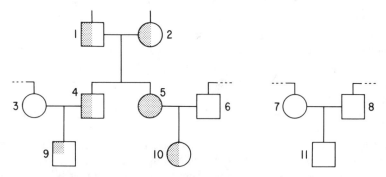

Fig. 2.3. Hypothetical pedigree diagram, applicable to any outcrossed diploid organism with separate sexes. Males are represented by squares, females by circles, mating by horizontal lines connecting males and females, offspring by verticals from the horizontals. Shading shows the proportion of the genes in each individual that are necessarily identical with those in 5. Her evolutionary success is measured by the extent to which she passes these genes to future generations. So the success of any other individual is proportionately equivalent to 5's success to the extent shown by the shading (coefficient of relationship to 5). Unshaded regions represent random samples of genes from the gene pool. Their transmission is normally considered irrelevant to 5's fitness, although it would be slightly detrimental in a small population.

selection among such groups is a substitute for assessment of inclusive fitness for their members. This substitution would be possible only in the unlikely event of complete uniformity of kinship within each group and negligible relationship between members of different groups. A possible example would be a bird population of seasonally monogamous females with no paternal assistance in the care of offspring and no possibility of brood parasitism. The groups would then be composed entirely of individuals with coefficients of relationship to each other of 0.5. If females dispersed from natal areas before breeding, there could be very nearly a 0.0 relationship between any two neighboring mothers. The effects of kin selection could then be subsumed under those of kin-group selection (Michod's (1982) *family-structured model*) for as long as the groups persist.

I suspect that Michod's (1982) requirements are never really met, and seldom closely approximated. Kin selection is a broader and more versatile concept than kin-group selection, as is clear from Michod's comparison. In mammals and some birds females tend to be philopatric so that neighboring adults, even when they do not form persistent kin groups, will often have recent genealogical connections. In both birds and mammals, broods of mixed paternity may be produced. This raises the possibility of the evolution of kin-recognition mechanisms and discrimination between full-sib and half-sib nest mates. Most birds also have paternal care and extra-pair copulation, and many are subject to within-species brood parasitism. These factors make conflict inevitable between the unrelated parents, and make it important, but difficult, for a male to assure that the young in his nest are his own, not merely his mate's, and for both parents to avoid fostering unrelated parasites. There would be selection for within-group kinship discrimination by both adults and nestlings.

It may be that much of the selection taking place in a woodland bird population may arise from the varying success of families. It is also true that a complete understanding of evolution in such a population would need to go beyond selection between family groups. It would require consideration of varying kinship within the groups, and perhaps of kinship among competitors. This conclusion often applies with particular force in the sociobiology of plants. A nut on a nut tree at first sight might appear to be a part of the tree, and on closer consideration an individual in its own right. Still closer attention shows it to be a composite of several genetically distinct individuals of varying kinship, each of which must be selected to optimize its interactions with each of the others. Queller's (1989) 'Inclusive fitness in a nutshell' is a provocative discussion of the conceptual challenge presented by any fruit or even a single seed.

The idea of kin selection assumes that organisms can evolve kinship-modulated mechanisms for dealing with other members of their species. Since kinship itself is not a perceivable attribute, selection must make use of indirect evidence, often something obvious, like propinquity at a certain stage of the life cycle. Two newly hatched birds in a nest will have some particular average degree of relationship. This average value will depend on the prevalent levels of inbreeding in the population, frequency and pattern of multiple paternity for single clutches, and frequency of brood parasitism. If the average relationship between nest mates is 0.4, and if propinquity is the only useful cue, a nestling is expected to treat another nestling as if its welfare were 0.4 as important as its own.

In recent years it has become apparent that organisms, in their assessment of kinship, are by no means limited to such information as the fact of sharing a nest. It is entirely possible that nest mates will perceive differences between full and half sibs and fostered parasites and modify their behavior on the basis of this information. They may make use of many kinds of comparisons of close associates with each other and with themselves. Mechanisms of kinship assessment and their role in the behavioral ecology of natural populations have been ably reviewed: Blaustein *et al.* (1988), Cockburn (1988, pp. 65–9), Fletcher and Michener (1987), Hepper (1986), and Waldman (1988).

2.4 Effects of inbreeding and asexual reproduction

The inbreeding that takes place in most wild animal and plant populations would not affect the arguments advanced above on the primacy of selection at the level of the gene in the codical domain. Even in plants that normally self-fertilize, or in animals that normally reproduce by brother–sister mating, occasional use of pollen from another individual, or occasional mating with a half sib or aunt or uncle, would imply some recombination of genes from previously separate lineages. In the long run this means that the genes that now form a particular genotype will go their separate ways and be separately selected on the basis of their long-term average phenotypic effects in the genotypes in which they occur. The genetic environment of such genes would be fine grained. Their history could be represented by dendrograms such as Fig. 2.1C, and selection in the codical domain would still be at the level of the gene.

In a small minority of plants and an even smaller minority of animals, such as the killifish *Rivulus marmoratus* (Harrington 1975; W. P. Davis 1988), reproduction is entirely by selfing. Their genes cease to recombine in a few generations. Different lines of descent may have different genes, but each line becomes rapidly homozygous for whatever genes it contains.

These entire genotypes become the codices that persist. Genotypic selection could be depicted by Fig. 2.1C, and it is at the level of genotypes that the selection of information would take place. There is no population in the Mendelian sense, and no gene pool. Exclusively brother–sister mating gives complete homozygosity more slowly than selfing, but the result is the same. Selection can act only among the different homozygous genotypes. The fate of each component gene is determined by the reproductive success of the somata specified by the genotype of which it is a part.

Asexual reproduction, whether vegetative or parthenogenetic (Suomalainen *et al.* 1987), likewise locks each gene into a lineage of persistent genotypes, although the cytogenetic machinery and genotypic consequences are totally different. All heterozygosity is retained in asexual lineages, and it can be expected to increase from occasional mutation. Here also, selection in the codical domain must be entirely at the level of the genotype rather than the gene, and this genotypic selection can be modeled realistically by a dendrogram (Fig. 2.1C).

The situation is more complicated and interesting if, as is common in most plant and many animal groups, reproduction is usually asexual but occasionally outcrossed sexual. Ultimately some quantification of *usually* and *occasionally* must be achieved for rigorous understanding, but for most examples the answer to the important question is clear. On an evolutionary time scale, recombination takes place often enough to give each gene a long-term historical independence of every other gene. However many physiologically separate individuals ultimately derive from a single zygote, they should be regarded as constituting a single individual in an evolutionary model (Harper 1977; Janzen 1977). Fig. 2.2 can be made to describe the situation merely by substituting *somata* for *soma*. Selection in the codical domain is at the level of the gene, and is based on its average phenotypic effects on the individual somata in which it finds itself.

3

Clade selection and macroevolution

Chapter 2 proposed that genes and, sometimes, genotypes of inbred or asexual lineages meet requirements for the precise long-term proliferation needed by units of selection in the codical domain. If the same can be true of gene pools, they must also be considered as likely codices for the natural selection of information. A gene pool and all its descendant gene pools constitute a clade, the usual sort of entity represented by diagrams such as Fig. 2.1A. Discussions of higher levels of selection are not always clear as to whether collections of gene pools or of organisms living at a particular moment are being considered. My use of *gene* and *clade* as strictly codical concepts demands other terms for their physical manifestations (Table 3.1). My consistent use of the term *clade selection* always implies a parallel *phylad selection* in the material domain. This is reasonable because of the simple one-to-one relation between the codex

Table 3.1 Levels of selection in the codical domain and the physical manifestation of each. The level *genotype* would not be a level of selection when reproduction is sexual.

Codex level		Physical manifestation
Gene		DNA molecule and its products and later phenotypic effects (unit character, or alteration of some quantitative or numerical character)
Genotype		Soma(ta). Different genotypes may produce different somatic phenotypes
Clade	Gene pool	Mendelian population
	Group of related gene pools	Phylad,* with or without formal taxonomic recognition

*A term used in much the same sense by Carson (1989).

(gene pool or monophyletic group thereof) and its material manifestation (population or phylad). There is the same simplicity for the selection of an asexual genotype and its clone. Selection at the level of the gene is complicated by its possible dependence on a number of levels above and within the individual soma (as discussed in Chapter 4).

3.1 Requirements for selection among gene pools and clades

Discussions of higher levels of selection have been obfuscated in recent years by the term *species selection* (Stanley 1979; S. J. Gould, 1982*b*; Vrba 1984) which implies that there is something special, for natural selection, about the species level of classification. Another inappropriate term is *taxon selection* (Williams 1985), which implies that a clade must be taxonomically distinguished to be relevant to selection. If I am right that all levels of genealogical inclusiveness from local gene pools to classes and phyla may be subject to selection, the term *clade selection* used by Stearns (1986) is a great improvement.

One requirement for clade selection is that, if a population includes the ancestors of later individuals, this relationship must be important in determining the properties of the descendants. These properties must be recognizable to biologists for clades to be distinguished and clade selection studied. This means that rates of cladogenesis must be rapid compared to rates of change within a clade or to divergence between recently formed clades. There is a formal parallel to the requirement of multigenerational stability of genes. We must be able to determine their presence by phenotypic effects, and they must have consistent effects on fitness to be selected effectively. This stability requirement is undoubtedly met at many levels of clade inclusiveness. A hundred million years from now, a descendant of some present vertebrate will surely be recognizable as a vertebrate and distinguishable from some coeval descendant of a present arthropod. I would likewise assume that, a hundred thousand years from now, descendants of brown trout and rainbow trout will form two distinguishable phylads.

Absolute stability of either genes or gene pools is surely impossible, and if it happened would rule out all evolution of the gene pool or clade. Rare mutations provide the alternative alleles of genes within a gene pool, which can then be selected on the basis of their effects on individuals, or on interactors at higher or lower levels (Chapter 4). Divergent evolutionary changes in gene pools provide the variation on which clade selection can take place. These within-clade changes are the higher-level analogue of gene mutation. One gene pool may be in an environment that selects for high equilibrium levels of A and b, while another environment selects a related gene pool for high *a* and *B*. If these or other changes from selection or drift result in somewhat different ecological niches or reproductive processes, later coexistence of the mostly *AAbb* and mostly *aaBB* in the same environment may be possible and the two populations would have become specifically distinct. Selection may continue to select for the same divergent gene frequencies in the two gene pools, irrespective of what frequencies at the *a*- or *b*-locus would be best for long-term group survival. Diverging allelic frequencies among gene pools can thereby provide alternatives for selection at the higher level.

I presume that the dependence of clade selection on differences in fitness among clades is rather obvious. One clade must have a set of genes more likely than another's to lead to long-term clade survival. This means that the ratio of probabilities of cladogenesis and extinction is different in the two clades. It does not matter how the clades came to differ in fitness. It could have been from any form of selection or from drift. There need be no implication that evolution within each clade was in any way influenced by considerations of clade survival, any more than occurrence of a mutation need be influenced by its effects on individual fitness.

There are many further questions on the meaning and limits of clade selection. One issue is whether the populations that bear the gene pools need be in ecological competition with each other. I believe that this is not required, any more than individuals within a population need interact ecologically to be subject to individual selection. The reproductive success or failure of a soil arthropod, with an expected lifetime dispersal of a few meters, will hardly influence prospects for a conspecific a hundred meters away. But the descendants of these two individuals might compete, and genes passed on by one may ultimately prevail over those passed on by the other. Selective elimination of one and survival of the other a hundred meters away is individual selection as long as the two arthropods can be assigned to the same population and their genes to the same gene pool. The ultimate prize in the competition between genes is representation in the gene pool.

In the same way, two gene pools in allopatry can be subject to natural selection if, as must always be true, their descendants might be alternatives for representation in the biota. Suppose a climatic change causes the brown trout of the upper Rhine to die out but lets the brown trout of the upper Danube survive. Suppose further that the difference in fate is attributable to some difference in gene frequency that causes a difference in vulnerability to the change. That is surely clade selection. The ultimate prize for which all clades are in competition is representation in the biota. If the survival of the left clade and extinction of the right in Fig. 2.1A is attributable to differences in genetic composition rather than chance, it exemplifies clade selection, regardless of the geographic distribution of either clade. The lack of dependence of clade selection on clade sympatry is implied by the many group-selection models (e.g. Slatkin 1987; D. S. Wilson 1987) that operate by differential ability to colonize new habitats. Clade selection is thus a broader concept than ecological succession, which requires local competitive replacement of one population by another.

There are also questions on requirements for size and taxonomic rank and duration of clades competing for representation in the biota. I would maintain that the first two are irrelevant. Clades of any size and degree of difference are subject in principle to natural selection. Selection among clades capable of later fusion with others (Fig. 2.1B) may possibly have important evolutionary effects (Bonhomme 1986; Dial and Marzluff 1990). There is a duration requirement, because a clade must be a line of descent and last long enough for there to be some turnover of generations. Some short-term examples of group selection, such as the *haystack* model of local buildup of a population followed by dispersal to new habitats (Maynard Smith 1976; D. S. Wilson 1987; Taylor and Wilson 1988) may formally qualify as clade selection. Within-generation trait-group selection does not. I discuss haystacks and trait groups in Chapter 4.

Clade selection is conceptually clearest when it deals with holophyletic sister groups, such as the African mammals discussed by Vrba (1989), but it need not be so restricted. It is legitimate to consider the steady replacement of a deuterostome phylum (Brachiopoda) by a proterostome class (Pelycepoda) as a possible example of clade selection (Vermeij 1987). Likewise a major taxon like the trilobites can be considered subject to unfavorable clade selection even without specifying which groups displaced it (Raup 1987).

My view of clade selection conforms to that of Van Valen (1988) but is contrary to some others. Maynard Smith (1984a), Sober (1984, pp. 258–61), and Vrba (1984, 1989) maintain that clade fitness (restricted to *population* or *species* fitness in many discussions) must result from emergent group properties that can not be ascribed to single individuals. One clade might be fitter than another from having a more productive sex ratio or more cooperative social system, but not from having greater disease resistance or better vision. I find this restriction unacceptable for various conceptual and practical reasons. One problem is that I would look for group-level adaptations, such as a thoroughly cooperative social system, as a product of clade selection and evidence of its operation in the past. To make it a prior condition for the occurrence of clade selection is like starting with a singing bird in an evolutionary explanation for bird song.

Another difficulty is the unlikelihood of achieving any general agreement on what is or is not an emergent group property. Sex ratio by definition is a property of a group, but it can result largely from individual allocation strategies in the Hymenoptera and other organisms. Disease resistance seems like a matter of individual fitness, but it may vary in relation to such group properties as herd size. Levels of inbreeding, population viscosity, and related group properties relate to magnitudes of individual dispersal. Every group property arises from characters developed by its members, and every individual character has its collective effects. Surely any listing of biological properties as at individual or higher levels would be largely arbitrary. The answer to all these difficulties must be Lloyd's (1988, pp. 101–7) idea that higher levels of selection depend, not on emergent characters, but on any and all emergent fitnesses. If any process operating at a lower level, such as selection of genes within gene pools, results in differences in fitness among gene pools, it has produced the kind of fitness variation required for a higher level of selection.

A purely practical consideration is the paucity of information about characters, individual or collective, in organisms known only as fossils. Suppose that, during the Cretaceous, one family of snails started with one genus and ended it with a hundred, while another went from a hundred to zero. Appropriate additional information and statistical tests might then be used to show that the first family had consistently higher rates of cladogenesis and lower rates of extinction than the second. It is surely of value here to recognize that the first family had, in an important sense, gene pools superior to those of the second, even though it is most unlikely that we will ever know whether it was from coding for more cooperative foraging or for broader temperature tolerance.

3.2 Selection of genes vs. selection of gene pools

The history of gene pools in a clade and of genes in a gene pool can both be realistically modeled by dendrograms (Fig. 2.1A, C). This means that many of the same concepts and forms of reasoning are applicable to both. There are also some noteworthy differences, one of them conspicuously illustrated in the figures: A history of origins and histories and disappearances of genes must be punctuational, because changes in genes are always instantaneous and quantal. A gene mutation implies no less than a base-pair substitution in the DNA that codes it. This would be a four-bit change in information, and some mutations must affect many base pairs. This kind of change is shown (Fig. 2.1C) by instantaneous lateral shifts of varying distances. For the moment I will anticipate later discussion (Chapters 8 and 9) and assume that clade histories are gradualist (Fig. 2.1A, B). Different gene pools gradually diverge from each other, often blending again if the initial divergence is from temporary allopatry (Fig. 2.1B). Once intrinsic reproductive isolation is evolved, the separate lineages can thereafter have genetically independent histories even in sympatry, and will diverge indefinitely wherever they may be.

Another important comparison is that of time scale. It has often been pointed out that individuals replace each other more rapidly than whole populations. It seems to follow that selection at the individual level, the main mechanism of selection at the genic level, must be more powerful than at the population level. This is an unfair comparison in its implications for the codical domain. The comparison should be between rates of turnover of genes and gene pools. This is clear from use of a dendrogram to model both processes. The microevolutionary analogue of the extinction of a clade is the extinction of a gene, not the death of an individual. The selective death of an *AA* individual need not eliminate *A* from a gene pool. If it does it would probably have been eliminated by drift anyway. I am unaware of any serious attempt to compare rates of origin and extinction of genes with those of gene pools.

Another traditional reason for believing that selection within is more powerful than selection between gene pools is the sample-size comparison and the implication that random processes must be more important at the clade level. The number of competing branches in a clade must often be far fewer than the number of simultaneous genotypes bearing a specific gene from a single gene pool. Here it is indeed the number of individuals that is relevant.

As a hypothetical example, suppose that the larvae of closely related moths exploit different host plants. One feeds on red oak, another on white oak, another on black oak, etc. Many millennia later some of these are extinct, and others have diversified; perhaps the one that fed on white oak is gone and that which fed on black oak is still there and has given rise to a descendant form that feeds mainly on pin oak. Can we safely conclude that the clade that fed on black oak has been favored by selection? Surely not. Successes and failures among entities each represented by a sample of one is not valid evidence for any kind of bias. What is needed is significant evidence that specializing on white oak is more likely to be followed by extinction and specializing on black oak more likely to lead to phylogenetic success. Even if white-oak specialization is always phylogenetically lethal and black-oak specialization always the start of an adaptive radiation, a sample of six clades of each type would be needed to establish the conclusion at the 0.05 confidence level.

By contrast, suppose, for a Mendelian population of moderate size (e.g. 1000 breeding adults every generation), that *a* changes from a frequency of 0.2 to 0.9 and *A* from 0.8 to 0.1 in 100 generations. This looks like convincing evidence for the superiority of *a* (population geneticists can no doubt think of some qualifications to such a conclusion). In its passage through 100 generations, *a* was always represented hundreds of times. However variable its phenotypic effects might have been in the various genotypes and environments in which it occurred, it seems to have had some mean effect more favorable to reproductive success than its allele *A*.

The analogy between gene and gene pool fails in some other respects. In traditional genetic modeling, two genes in a gene pool are either allelic or not, and it is only allelic genes that are clearly in competition with each other. If at the a-locus in a diploid population of 1000 breeding adults there are two alleles, *A* and *a*, there is, in effect, a population size approaching 2000 for statistical consideration of events at that locus. If there are four times as many *A*s as *a*s, there are about 1600 of one kind of contender and 400 of the other. Even a rather slight net advantage of one over the other will give an almost deterministic shift in the relative frequencies of the two alleles. Only if the two are of almost identical fitness, or if there were an extremely small number of one allele, could the generation-to-generation sampling error known as *genetic drift* be more important than selection in determining future allelic frequencies.

In clade selection there is no clear form of allelism. A given population may be in ecological competition with many others, and all clades are in competition for a place in the biota. Questions of sample size depend on what question is asked. If one asks how important selection among its gene pools is likely to have been in the evolution of the genus *Homo*, the answer will depend on one's guess as to how many species and local populations of *Homo* there ordinarily were since the lineage first broke from that leading to *Pan*. If only a few clades were present at any given time, great differences in clade fitness would be necessary for clade selection to be more important than chance in determining the future composition of the genus. If there were often hundreds of separate gene pools in our genus in tropical Africa over the last few million years, moderate fitness differences could have made clade selection quite important.

A somewhat different question might get quite a different answer. One might be interested mainly in the role of clade selection in determining species-level composition of the genus *Homo*. The sample size would then be however many species-level clades were in existence at any one time. I presume that there never were more than a few synchronous species in this genus. So for this explicitly species-selection question the sample size would be small and chance far more likely than selection to be important.

It might seem that there is an important difference in the sources of variation on which selection acts on genes and clades. Gene mutation is a random alteration of a message that has been subject to selection for maximum fitness during the whole history of the lineage. It is expected to reduce fitness far more often than it is to increase it. The microevolutionary changes that cause diversity among gene pools may be largely the result of favorable selection of individuals. This kind of change may seem more likely to be favorable at other levels than would purely random changes. For this reason I think it more likely that a selection-generated divergence of a clade from its ancestral gene pool is more likely to be favored in clade selection than a random mutation is to be favored within a gene pool. Gene pool divergence driven by drift is more analogous to random mutation. The difference should not be overrated. Just as gene mutation occurs without regard to its effects on the individual fitness of its bearer, changes in gene frequencies, from either selection or drift, occur without regard to effects on the persistence of the gene pool. In this important sense, selection at both gene and clade levels works on random variation.

I suggested on page 18 that the traditional calculation of relative allelic frequency may not be a perfect measure of success at the genic level. Difficulties in evaluating clade success are far greater. What really ought to be measured is the extent to which a monophyletic group of gene pools imposes pattern on matter. A count of individuals would not be correct, because large and active individuals, rapidly diverting resources into the service of their own reproductive interests, would be imposing more pattern on matter than would small inactive ones. Counts of gene pools or of subordinate clades (e.g. species in competing genera) would not be right, because the gene pools of large populations impose more pattern than those of small ones. Counts of individuals or of populations, and various other simple measures, would be correlated with the theoretically important one, and can therefore serve as approximations. The choice could be made on the basis of convenience and repeatability. I will generally assume that numbers of gene pools or of subordinate clades is an acceptable metric of clade success. It indicates the number of contenders subject to selection, and permits clade success to be shown by, and read from, a dendrogram.

I think the logical similarities of selection at genic and clade levels more impressive than the differences. Selection of both genes and gene pools results from biases in survival and replication of competing codices (Fig. 2.1A, C) that are intrinsically capable of indefinite persistence and proliferation (Dretske's principle). A gene can be favorably selected in a gene pool on the basis of any and all advantages over competing genes. A gene pool can be favorably selected because of any and all advantages it has over competing gene pools. Neither process must be a strictly zero-sum game, but one entity's success must often imply another's failure.

3.3 How important is clade selection?

In my opinion the recognized microevolutionary processes that form the heart of the neoDarwinian synthesis are an adequate description of the evolution taking place in any Mendelian population. In particular, the natural selection of alternative alleles, acting largely independently at each locus, is the only force tending to maintain or improve adaptations shown by the ephemeral organisms formed by the ephemeral genotypes. If one could look back through the evolution of our own or any other sexually reproducing species, back to well before the Cambrian, no other fitness-enhancing process of any importance would be found.

Having taken that position, I must also take another. The microevolutionary process that adequately describes evolution in a population is an utterly inadequate account of the evolution of the Earth's biota. It is inadequate because the evolution of the biota is more than the mutational origin and subsequent survival or extinction of genes in gene pools. Biotic evolution is also the cladogenetic origin and subsequent survival and extinction of gene pools in the biota.

This position is not equivalent to a belief in the efficacy of clade selection. A purely random flux of cladogenesis and extinction would inevitably have important macroevolutionary consequences. If the hominid clade had gone extinct shortly after its origin a few million years ago, it would have important consequences for today's biota no matter what caused the extinction. It might have been from clade selection against bipedalism or from a local volcanic catastrophe where the clade was getting started. I would call this catastrophe chance extinction, but chance in this context would not mean that clade origins and extinctions happen without understandable causes. It merely means a freedom from any consistent and demonstrable bias. S. J. Gould (1989 and earlier) has consistently emphasized the importance of chance extinction in the evolution of the Earth's biota.

Remarkably little effort has gone into searching for consistent and demonstrable biases in phylogenetic history. Van Valen's (1975) is a pioneering example of such a search, routinely neglected by later workers. It provides convincing examples of clade selection, one that was already widely known (or at least widely believed) and two others that had not yet been recognized. The previously known kind was selection for the retention of sexual reproduction (recently discussed by Nunney (1989)).

It occasionally happens that female hybrids are formed that produce diploid eggs as a result of the disruption of meiosis in hybrid oogenesis. Sometimes such eggs may develop into viable and fertile daughter females. In this way a new species may be formed, one that consists of a single clone of genetically identical females. Repetition of such events produces parthenogenetic species with clonal diversity. Cyclically parthenogenetic species can also produce asexual ones by the loss, from genetic or environmental change, of the sexual process. Organisms that can reproduce vegetatively may be purely asexual species where environmental conditions interfere with sexual reproduction. Van Valen showed that species that originate in this way, or in any other that results in exclusively asexual reproduction, are always taxonomically isolated. It apparently never happens that such species give rise to a group of related asexual species (for exceptions see Chapter 10). Selection consistently favors sexual clades over asexual clones or groups of clones.

One of Van Valen's new findings is that mammalian clades with large body size have both an elevated extinction rate and an elevated rate of production of new species, but that the extinction effect is stronger, so that clades of large-size mammals tend to die out. This clade selection against large mammals is balanced by a tendency for small mammals to evolve a larger body size. Van Valen's third finding was an opposite effect for the Foraminifera. There is net clade selection against smaller forams, but this effect is balanced by large species often becoming small. An important picture emerges: if it were not for clade selection, asexual organisms would be more common than they are; the present mammalian fauna would consist to a greater extent of large-bodied forms; and the foraminiferan fauna would include a larger proportion of small forms. Van Valen's work dealt only with these few examples, but it certainly suggests that clade selection may be generally important in macroevolution and that further searches for comparable effects might be rewarding.

Such a search, with similar results, was recently reported by Mitter *et al.* (1988). They found evidence, among several orders of insects, that clades that switch to eating angiosperms from any other diet have greater phylogenetic diversification than is shown by sister groups that retain the original diet. They were not able to determine whether the greater diversity resulted from increased cladogenesis or decreased extinction. Other studies have produced negative or equivocal evidence. Neither Hoffman's (1986) nor Benton's (1987) surveys of the fossil record found any evidence for macroevolutionary effects of clade selection. They conclude that a purely random process of clade branching and extinction could have produced the observed macroevolutionary patterns. Carroll (1988, pp. 576–8) found little indication of clade selection in the history of vertebrates. Likewise Raup (1987) found only random waxings and wanings among invertebrate groups during the Ordovician.

Raup (1986) looked at extinction throughout the Phanerozoic, without considering rates of clade origin, and gave special attention to the existence and importance of occasional episodes of much higher than average extinction rates. He concluded that extinctions are selective, especially the occasional mass extinctions, but that the selectivity may counter what seem to be general trends. As an example he pointed out that a certain range of increase in ionizing radiation could cause the disappearance of mammals and birds, but have only minor effects on insects and plants. Jablonski (1986*a*,*b*) also examined the record of extinctions and concurred with Raup on its selectivity and on differences in selectivity between normal and catastrophic extinctions.

As noted on page 27, it should be much easier to show that clade selection operates than to show why. Van Valen showed clade selection against mammals of large size but perhaps not selection against large size. When large species go extinct and small ones persist, there is no assurance that size itself caused the difference. Size would be an indirect cause if, for instance, selection was actually against small populations. Large body size would often result in a small population, vulnerable to chance extinction by numerical fluctuation (Pimm *et al.* 1988). Similarly in the study by Mitter *et al.* (1988), the cause of the increased success of phytophagous clades is not obvious. They suggest that phytophagy may favor insect diversification because there are so many different kinds of higher plants. This would increase cladogenesis by providing many alternative food sources to be exploited by diverging clades, and then might curtail extinction by reducing competition between closely related sympatric forms. Of course, the observation that a feature is favored as a result of clade selection need not imply that that feature was even indirectly responsible. There could be just a chance association between the observed feature and the one really responsible. Sober (1984) makes this point in his discussion of the difference between selection *of* a feature and selection *for* it.

I would tentatively favor accepting body size and phytophagy as at least indirectly responsible for the selection in these two studies. Both dealt with a great taxonomic and ecological diversity of organisms, and this reduces the likelihood of their all having some feature in common, other than the one studied, that would explain the observations. Van Valen's mammals belonged to many separate orders and lived in habitats from the pelagic marine to montane forests. It seems unlikely that there would be some factor in common to all these groups other than size, or effects of size, that could cause the different rates of extinction and cladogenesis. Likewise in the study by Mitter *et al.*, many species in several insect orders were tabulated. Their data must include a large number of largely independent associations between diet and diversity. It is unlikely that all the studied phytophagous groups were consistently similar in some crucial character other than phytophagy and its effects.

A similar problem arises in attributing selection to a character within a population, as Sober (1984, pp. 98–100) emphasized. The demonstration that some allozyme heterozygote has higher than average fitness is hardly proof that the enzymatic phenotype is the cause. It could be another effect of some genes in linkage disequilibrium with the allozyme locus. Actually, linkage disequilibria in natural populations are usually undetectable and seldom more than a few per cent (Spiess 1977, Chapter 17; Maynard Smith 1989, Chapter 5). The enzyme itself remains the most likely explanation of increased fitness (Chapter 5). Features may remain associated in phylogeny for purely historical, rather than functional reasons, and make it hazardous to infer favorable selection of some feature merely because the clade that bears it was favorably selected. Such inferences gain credence as the number and diversity of the studied clades increases.

The main problem with assessing the operation and importance of clade selection is that of distinguishing it from randomly varying rates of branching and extinction (Hoffman 1987). The problem is analogous to that of distinguishing drift from selection within the gene pool of a Mendelian population. Within a gene pool there will be successful and unsuccessful genes that can be depicted as genealogies on a dendrogram (Fig.2.1). Within a biota the separate gene pools can be represented the same way. For either kind of codex, the demonstration of selection is critically dependent on sample size. As noted on page 28, much larger samples are often available for competing genes in a gene pool than for clades in a taxon or ecological community. This factor increases the importance of chance in macroevolution and the difficulty of demonstrating clade selection in phylogenetic history.

3.4 Characters likely to be important in clade selection

The idea of progress in evolution antedates Darwin but has lost favor since his time (Provine 1989). In a less grandiose form it may be recognized in Brown's (1958) concept of *general adaptation*, a feature that may give one clade an advantage over others and make it the start of an adaptive radiation of organisms with that feature. Among Brown's examples were mechanisms for outcrossing and of individual dispersal. He found evidence that abundant and widespread species, which he believed to have the primary potential for evolutionary success, were usually outcrossed and widely dispersed. Related species of restricted distribution were more likely to be inbred and philopatric. In my terms, then, outcrossing and wide dispersal of individuals at some stage of the life history seem to be favored by clade selection.

One of the most widely recognized examples of a new and improved group replacing a more primitive competitor is the ascendancy, in many habitats, of angiosperms over gymnosperms. This apparent adaptive superiority of the flowering plants is traditionally attributed to their methods of pollination and seed nurture. Bond (1989) agrees that the angiosperms are indeed superior in the more productive habitats, but for reasons unrelated to reproductive processes. He argues that the angiosperms have a competitive edge, especially in the seedling stage, because their more efficient vascular systems permit more rapid growth.

By far the most conspicuously successful order of birds is the Passeriformes, and various ornithologists have proposed various reasons for this, special passerine features that enabled them to spread, diversify, and avoid extinction. Raikow (1986) re-examined the arguments and concluded that no single feature of passerines can account for their success, which he believes to be largely an artifact of ornithological classification practices. This dispute is still unsettled (Slowinski and Guyer 1989). Recently Stiassny and Jensen (1987) suggested that the perciform suborder that includes damselfish and wrasses owes its success (ca. 1800 species) to a structural modification that gives special capability to the action of the jaws. If valid, this example illustrates Brown's (1958) concept of general adaptation and Liem's (1990) concept of *key evolutionary innovation*.

I have no reason to doubt that various new adaptations evolved in certain lineages have facilitated their subsequent evolutionary proliferation. I seriously doubt that there is any way of establishing this proposition, because there is no way of knowing how any such group would have fared without its supposedly special advantage. This and other objections have been raised by Slowinski and Guyer (1989), Cracraft (1990) and Jensen (1990). I believe that the clade selection idea will be much more useful for simple categories of character such as large size, phytophagy, and absence of recombination. These are characters that can be evolved independently by many different groups and allow appropriate statistical tests to be made.

Only the barest beginnings have been made in searching the fossil record for evidence of clade selection. The record can be searched for statistically significant trends in diversity and abundance of particular clades, such as trilobites, brachiopods, and pelecypods (Raup 1987; Vermeij 1987). It can also be searched for consistent selection on certain characters, such as body size (Van Valen 1975). This kind of search could be made on any group of animals or on fossil plants, using such features as stem diameter as a measure of body size. Sizes of parts (leaves, seeds, pollen grains) could also be surveyed.

The consistency of clade selection against exclusively asexual reproduction may well be the best established and most widely applicable principle of macroevolution. It suggests that evolutionary changes that compromise sexuality in any way, even though some recombination can still take place, may also be selected against. Various sorts of compromises of the sexual process, such as loss of heterostyly or other assurances of outbreeding are often evolved but are unlikely to be reversed by selection within a population (Barrett 1989). The prevalence of outcrossing in flowering plants may result from selection for selfing within gene pools often being countered by selection for outcrossing between gene pools. Exclusively outcrossed plants may thus be favored by clade selection over those that rely partly on selfing or apomixis, those with less over those with more use of resources for vegetative spread, and cyclically parthenogenetic animals with frequent sexual phases over those with infrequent.

A priori one might imagine that populations that use resources efficiently for their own maintenance and reproduction would be favored over the less efficient. Evidence for inefficiency could be seen in any expenditures on competition between members of the population. Such competition could be overt, and centered on limiting resources, or more subtle, with social status as its immediate object (West-Eberhard 1983). The most obvious examples relate to competition for mates. Marked sexual dimorphism or other evidence of strong sexual selection would be evidence for an inefficient use of resources by the population. The same could be said for the related factor of paternal investment. Populations in which males help to raise the young ought to win out over those in which only the females show parental care. I envision the possibility that at any given moment the natural selection of individuals in most populations is leading to increased polygyny, more wasteful strife among males, and decreasing paternal investment in young. This individual selection is countered by clade selection, so that the minority of current populations that has the opposite trends will have more descendants in the biota a million years from now.

Evolutionary changes that are irreversible (Bull and Charnov 1985), or merely proceed in one direction more easily than another, such as those discussed above, or the loss of feeding larval stages in marine animals (Strathmann 1978), must often be countered by clade selection. Other possibilities discussed by Bull and Charnov, such as sex determination by heteromorphic chromosomes, haplodiploid sex determination, and polyploidy, imply that the continued occurrence of environmental sex determination, and of modest chromosome numbers, results from favorable clade selection. A special use of the idea of opposed selection at genic and clade levels is made in Chapter 9.

I anticipate much greater attention to clade selection and other macroevolutionary topics in the near future. The few paragraphs above, and related discussion in Chapters 4 and 9, are hardly an adequate treatment of so complex and important a topic. I hope I have adequately shown that clade selection does operate, at as yet unknown levels of frequency and intensity. It must also be true that there is a strong stochastic element in phylogenetic success and failure, and that chance events can have major effects through all subsequent history (Chapter 6).

4

Levels of selection among interactors

Natural selection must always act on physical entities (interactors) that vary in aptitude for reproduction, either because they differ in the machinery of reproduction or in that of survival and resource capture on which reproduction depends. It is also necessary that there be what Darwin (1859, Chapter 4) called 'the strong principle of inheritance', so that events in the material domain can influence the codical record. Offspring must tend to resemble their own parents more than those of other offspring. Whenever these conditions are found there will be natural selection. Wherever they are found to great degree: inheritance strong enough, differences in fitness great enough, competing alternatives numerous enough, selection may produce noteworthy cumulative effects.

To Darwin and most of his immediate and later followers, the physical entities of interest for the theory of natural selection were discrete individual organisms. This restricted range of attention has never been logically defensible, especially not since Lewontin's (1970) lucid examination of the problem and of much relevant literature more than 20 years ago. Interactors can conceivably be selected at levels from molecule to ecosystem, and there has been helpful recent progress on this levels-of-selection question (Sober 1984, Chapters 7–9; Brandon 1988, 1990; Damuth and Heisler 1988; Lloyd 1988).

4.1 Adaptation as the material effect of response to selection

Natural selection at any given material level, such as the individual soma, is expected to produce adaptation at that level, unless heritability is zero or selection is overcome by random processes or selection at other levels. If no evidence of adaptation is found, it must be that selection is absent or ineffective at that level. Relationships among such concepts as adaptation, selection, fitness, and gene-frequency change have long been matters of everyday significance in evolutionary biology. More recently they have gotten detailed attention from philosophers, whose writings are now worth careful attention from biologists. Brandon (1990) is a good example and introduction to other recent contributions.

I doubt that I could provide any general discussion of these topics that would be an improvement over some others that are available, such as Ayala's (1968) or Lloyd's (1988). I will confine attention to the recent controversy on how adaptation ought to be identified, a matter implied in many discussions and clearly confronted by Symons (1987, 1989), Wade (1987), Wade and Kalisz (1990), Mitchell-Olds and Shaw (1990), and Thornhill (1990). An obvious approach is simply to show that a proposed adaptation does indeed contribute to fitness. Wing clipping a bird impairs flight, an observable decrease in adaptive performance. The obvious conclusion is that the wing, not only by its presence, but in its normal size and shape, is a flight adaptation. Some studies may go far beyond some measure of immediate performance, and consider effects on net reproductive success and gene frequencies, over a whole life cycle, of natural or experimentally imposed differences between individuals, e.g. Cooke (1988) or several contributions in Clutton-Brock (1988).

The study of variation in performance has great value, but it is not a reliable way to demonstrate adaptation. Its inadequacy arises when findings are negative, as must often happen in studies of human adaptations. We expect organisms, human or otherwise, to use resources adaptively to increase the likelihood of reproductive success. Suppose we find otherwise, for instance that the rich have lower net reproductive success than the poor in some society (Vining 1986). Surely such observations need not lead us to deny the premise of human resource-use adaptation. We might be more inclined to suspect that wealthier individuals make more use of contraceptive technology and that this is a biologically abnormal problem that adaptations perfected in the Stone Age would not be expected to solve (Dawkins 1986; Symons 1986). Thus the observations could be interpreted in two ways: the proposed adaptation does not exist; or it exists but the data show a restriction on its *compass* (robustness over environmental variation).

Of course it often happens that putative human adaptations can be shown to augment performance. Hawkes *et al.* (1985) found that hunting behavior and meat distribution in a Paraguayan Indian tribe conformed to models of optimal foraging and resource allocation, despite many environmental abnormalities, and whether the hunters used archery or shotguns. Their study showed that an example of human decision-making on the acquisition and distribution of resources is adaptive and also that it has a compass broader than the conditions under which it evolved. If the findings had been in conflict with the model, they could have been explained by proposing either that the decision making is not adaptive, or that the adaptation fails in the abnormal environment of modern Paraguay.

Studies of wild populations must often face the same problem. Bluegill sunfish nest in colonies, with dominant males appropriating positions in the interior so that subordinates are forced to accept peripheral sites. Gross and MacMillan (1981) and Dominey (1983) proposed that preference for interior sites is an adaptation against egg predators, supposedly a greater hazard for nests on the periphery. Their observations confirmed this supposition, but what if they had not? Suppose the seemingly natural environment actually lacked a normally abundant catfish, or other egg predator. It might also be that eggs in interior nests are slightly more vulnerable to some pathogen so that peripheral nests could actually have a higher mean success rate when certain predators are absent. These observations would have shown that central nesting can be a net liability in the absence of egg predators. It would not be evidence against the proposed adaptation.

Even in truly natural environments, genuine adaptations may often fail to improve performance, because conditions vary in both time and space. A supposed adaptation that fails to augment fitness in 1991 may do so in 1992, or in another part of the forest. A gene pool is an imperfect record of a running average of selection pressures over a long period of time in an area often much larger than individual dispersal distances. It would be rash to assume that any brief and local biological study can yield reliable data on long-term directions and intensities of natural selection. Good examples of the variability of selection over a number of years are provided by Gibbs and Grant (1987) and by Pease and Bull (1990).

The best evidence on historically normal patterns of natural selection must come from an examination of the interactors themselves. Adaptation is demonstrated by observed conformity to *a priori* design specifications. This is the main method used by Galen and Paley (Appendix) and recently advocated by Thornhill (1990). The hand is an adaptation for manipulation because it conforms in many ways to what an engineer would expect, *a priori*, of manipulative machinery; the eye is an optical instrument because it conforms to expectations for an optical instrument; dosage compensation in *Drosophila* is an adaptation for canalizing similar development in males and females because it acts precisely in many ways to overcome developmentally divergent effects of male–female differences in ploidy (Muller 1948). Thornhill's (1991) argument that the abdominal clamp of male scorpionflies is an adaptation for holding uncooperative females is another good example of the same kind of reasoning.

Unfortunately those who wish to ascertain whether some attribute of an organism does or does not conform to design specifications are left largely to their own intuitions, with little help from established methodology. Standards for the design of experiments or for showing cause–effect relationships have long been matters of rigorous thought and well-vindicated scientific tradition, but they are only partly applicable to the demonstration of functional design. Experiments such as the wing clipping mentioned above can be undertaken to test understanding of machinery, with no regard for the effect on reproductive success. Wright and Cuthill's (1990) weighting of one parental bird to note effects on the behavior of the other are very much in this tradition, as was Power's (1990) moving of stones in a stream to test expectations on the foraging behavior of a catfish.

The comparative method (Chapter 7), of using the phylogenetic distribution of character states to establish their functional significance, has recently received serious thought, but relates to only part of the problem of functional interpretation. I suspect that this difficulty will be a matter of close attention from biological epistemologists in the near future. For the present I will merely assume the adequacy of the kinds of intuition that I share with Galen, Paley, Muller, and Thornhill.

4.2 Natural selection within organisms

Selection of molecular entities, such as the DNA that serves for the archival coding of the gene, has recently had considerable attention. An example is the selfish DNA of Doolittle and Sapienza (1980), which proliferates within a cell nucleus, perhaps at net cost to the host organism. Another is segregation distorters (Leigh 1977; Geliva 1987), genes or linked groups of genes that would be selected against on the basis of organismal effects, but which can counter this adverse selection by biasing the meiotic process so as to increase likelihood of inclusion in a zygote. Selection at the level of DNA molecules that code for such effects is in opposition to selection at the organismic level. Selection among whole organisms or phylads acts to suppress any such within-organism selfishness. A cytoplasmically inherited sex determiner could be favorably selected and quickly cause the extinction of the population in which it arises (Charnov 1982, p. 12).

The main theoretical challenge is currently not the reality of natural selection of these entities (transposable elements, outlaw genes, driving chromosomes, etc.), but rather with the general suppression of their short-term selfishness and the usual fairness of cell division and Mendelian heredity. This prevalent ruliness of DNA behavior means that selection at the organismic or higher levels must be much stronger, so that, in Leigh's (1977) words, 'It is as if we had to do with a parliament of genes, which so regulated itself as to prevent "cabals of a few" from conspiring for their own "selfish profit" at the expense of the "commonwealth".'

Many cytoplasmic organelles are observed to originate only by division of similar organelles. If they embody some of the information that determines their properties, they could be subject to natural selection. More rapidly proliferating mitochondria, for example, might replace the less prolific, regardless of effects on the cell they inhabit. Mitochondria, chloroplasts, and other cytoplasmic entities contain nucleic acids, and this strongly suggests that they have genotypes that direct their development. It is commonly believed that they originated as independent organisms competing for a common pool of resources. If so, their original induction into complex cooperative systems is an anomalous event, not to be expected on the basis of our usual understanding of the evolutionary process. The subsequent stability of these eukaryotic cell lineages through geologic time, despite potential disruption from selection among cellular components, presents an evolutionary problem that deserves detailed attention. ESS models and related ideas developed to aid understanding of cooperative behavior among animals may or may not be entirely applicable to the special circumstances of genetically independent life within a cell. Important steps towards the resolution of this problem have been taken by Eberhard (1980, 1990a) and Cosmides and Tooby (1981).

Buss (1987) has risen to the challenge of conflict resolution at the next higher level, that of the cell in multicellular organisms. These cells seldom overtly compete with each other for transmission in sexual or asexual reproduction. A privileged few are set aside early in development (in many animal groups) and the rest devote themselves entirely to furthering the reproductive interests of the special few. For plants and some animals the distinction between germ line and soma is made on a more continuous and stochastic basis, but ultimately most cells assume purely somatic functions that enable others to form propagules. As long as this cooperation among cells prevails, selection will normally operate between rather than within multicellular organisms.

The usual genetic homogeneity of the products of mitotic cell division makes altruism among cells much easier to maintain in evolution than cooperation among the originally independent components of the eukaryotic cell. I presume that the perfection of the mechanisms of cell division and Mendelian heredity antedated multicellularity. If so, the genetic identity (barring mutation) of the descendants of a one-cell propagule must have greatly facilitated the evolution of tissue specialization for somatic roles. Similarly, cooperation among clonally produced multicellular modules, such as those of a siphonophore or graptolite (Bates and Kirk 1985; Mackie 1986), is an expected development. If any somatic specialization by one zooid even slightly augments the reproduction of the colony, it will be favorably selected. This is kin selection with coefficients of relationship of 1.0, approximately true even with unusually high rates of mutation. The suppression of short-term selfishness among cells is of course not absolute, as the phenomenon of neoplasm dramatically attests. The need to suppress neoplastic cell division has probably resulted in some compromise of cellular mechanisms of healing and regneration (Miller, 1990; Williams and Nesse 1991).

4.3 Selection of individuals in populations

The term *individual* in abstract discussions in population genetics usually refers to a genotype, produced by a sampling of genomes from a gene pool. This is the usage I followed in Chapter 2. In other contexts it means a physiologically independent material entity, an interactor developed from genotypic instructions. For many of the organisms of greatest interest, genotypic and physical individuality coincide. Each physically identifiable individual has a unique diploid genotype formed by the union of two haploid gametes each with a genome sampled from the same gene pool.

For routinely cloning organisms this correspondence of genetic and physiological individuality disappears. Botanical custom refers to physiologically defined individuals as *ramets* and the whole collection of ramets of the same genotype as a *genet* (Harper 1977, pp. 24–6). I follow Grosberg (1988) and Hughes (1989) in extending this helpful terminology to cloning animals. A *Daphnia* zygote may produce a clone of millions of functionally separate somata (ramets) scattered through the plankton. Since they all have the same genotype, they are all parts of the same individual (genet) in evolutionary theory (Janzen 1977; Cook 1983). The fitness of that genet is the collective capability of the component ramets to contribute genes to the next generation of sexual progeny. Models of the haploid–diploid cycle of meiosis and fertilization are the same, regardless of whether the diploid interactor is physically single or multiple. The cloning of *Daphnia* has the effect of greatly increasing generation length (zygote-to-zygote interval) and the sexual fecundity of each genet (the somata of Fig. 2.2, p. 17).

It is surely at the level of the physically distinct individual that selection in most lineages has produced the most impressive results. Only individual somata really provide a wealth of conformity to *a priori* design specifications. They have been impressive to thoughtful observers throughout history (see Appendix), even though it is only in this century that we have known enough physics and chemistry to understand more than a few aspects of the machinery of life. Research continues to disclose the precision and complexity of the adaptive design of individuals from molecular to social levels of organization. Intricate enzymatic machinery is fine-tuned to organismal optima (Nevo 1986; Koehn & Hilbish 1987; Liles 1988), and we are coming more and more to appreciate the enormity of the difficulties that must be surmounted to keep even seemingly simple mechanisms working properly. An example is the operation of the cilia on the gills of a mussel, as discussed by Liles (1988). In his words, 'the cells and organs that make life possible had better be well designed, because the job of living is formidable.' Whole worlds of sensory and behavioral subtleties have been opened up; widespread kin-discrimination mechanisms (Chapter 2) and the advanced spatial conceptualization shown by bees (J. L. Gould 1986) are just two from an endless list of possible examples. I presume that there is little controversy on the efficacy of selection on such individual characters.

Buss (1987) makes the point that the evolution of life from its inception must have involved the encapsulation of the germ line in an ever expanding series of interactors and ever expanding levels within which competition is suppressed. The evolution of cooperation within a genet is facilitated by commonality of interest of all its ramets, whether they be cells or multicellular modules. Cooperation beyond that level is inevitably retarded by the conflicting interests of genetically different entities. Later sections of this chapter will argue that the evolution of functionally well organized systems of genetically heterogeneous elements has been rather rare, although when it has happened it has sometimes been phylogenetically quite important.

Associated conspecifics are inevitably competitors for resources, including mates. There is an unfortunate tendency to regard selection for effectiveness in sexual competition as something other than a kind of natural selection. If a habitat includes sexual competitors, that is an environmental feature that the individual must deal with no less than the problem of pathogens or food shortages. The special conceptual complexities of sexual selection theory actually characterize the broader category of selection for social status, which can be an important determiner of access to any kind of resource (West-Eberhard 1983). A dominant individual may deny a subordinate access to food as readily as to a mate. The concept of the social environment has recently shown its great potential for simplifying and clarifying thought on natural selection within populations. It plays a role, at least by implication, in any discussion of frequency-dependent selection or any use of ESS analysis (Chapter 5).

Selection at the level of the gene seems to imply a maximization of the ability of genotypes to get their component genes back into the gene pool. Since the fitness of genotypes depends on the phenotypes they produce, a maximization of phenotypic fitness may be expected. Unfortunately matters are not that straightforward (Dawkins 1982*a*, Chapter 10) in a world full of conflicting interests and pervasive deception. A vertebrate that devotes a significant proportion of its nutritive intake to the nourishment of a tapeworm is not an example of optimal resource allocation. A minnow that hunts insect larvae near a thicket that conceals a pickerel is not foraging optimally. The minnow's optimal behavior, staying well away from any pickerel of several times its own length, can be turned off by a pickerel that successfully claims to be absent when it is really there. By its mottled, countershaded coloration, stealthy movement, and hiding among plant growths, a pickerel says eloquently, 'I am not here, and it is perfectly safe for you minnows to forage nearby.' If a minnow is deceived, some crucial aspects of its habitat selection may be said to be produced by the pickerel's genes, not its own. This would illustrate the phenomenon of manipulation and of Dawkins' concept of the pickerel's *extended phenotype*. I have more to say about these matters in Chapter 5.

4.4 Trait-group selection

D. S. Wilson (1980, 1986) has proposed that even briefly interacting individuals comprise *trait-groups*, a higher-than-individual level of selection. One of Wilson's trait-group examples (1980, pp. 76–8) is winter flocks of song birds. Kinship among flock members is likely to be low and kin selection unlikely to have much effect on their interactions. It is nonetheless likely that flocks will vary in composition, perhaps more than expected of random association. Different flocks may have different average frequencies of aggression, emission of alarm calls, and other characters likely to affect the fitness of the flock as a whole. In the spring, when the birds go their separate ways and establish breeding territories, those from the more socially benign flocks can be expected to be more numerous, better fed, and better able to reproduce. The evolutionary result, if more successful flocks tend to preserve more cooperative individuals, should be an increase in those social traits that contribute to flock fitness.

Wilson's reasoning is clear and his examples convincing. Trait groups and related aspects of population structure must be important evolutionary factors, and future workers would do well to build upon the foundation laid by Wilson. It should also be borne in mind that trait-group selection is conceptually quite different from selection among populations or phylads. The most immediate contrast is that a trait group does not reproduce and has no persistent codex. A history of its success or failure is not recorded in the codical domain and can not be modelled by a dendrogram (Fig. 2.1). Even kin groups, perhaps the most important trait-group category, usually disappear as their individual members disperse and mate exogamously. The codices that record trait-group selection are genes, not gene pools.

It is also often true that problems analyzed by trait-group models can also be modeled by individual-selection models based on the realization that the individuals are selected in a complex environment that must be categorized in many ways. The relevant concept in relation to single-species trait groups would be the social environment. Other members of its own species are very much a part of the environment of an individual. It may participate in temporary associations with conspecifics, and there are likely to be costs and benefits of such association. There can be net benefits to group membership even for altruists that do not do as well as other members of the same group. These costs and benefits will influence the reproductive success of such an individual and thereby register themselves in the codical domain as adjustments of gene frequency and improve adaptation to the social environment. Wilson's trait-group selection can be modeled as equivalent to selection based on the success and failure of individuals as influenced by membership in trait groups.

This criticism of trait groups as a level of selection above the individual has been made in greater detail by others (Parker 1984; Nunney 1985*a*, *b*; Damuth and Heisler 1988; Nunney and Luck 1988). Another possible criticism is that, if trait groups were an effective level of selection, it ought to optimize the properties of trait groups. I will argue in Chapter 5 that interactions in trait groups are determined largely by frequency-dependent selection, which optimizes nothing at any level.

4.5 Selection of populations within species

The realization that selection can occur at phenotypic levels above that of individual organisms has a long history, ably reviewed by D. S. Wilson (1983). I take issue with Wilson on the importance generally accorded to group selection prior to 1960. My recollection is that the main use of the theory of natural selection in the early and middle years of this century was as an excuse to believe in evolution. Of course there was some use of the theory to explain adaptation, often in relation to climatic differences over a geographic range. Studies by Clausen *et al.* (1941, 1947) on plants that ranged from alpine to lowland habitats are a classic example. Others are the several climate-related rules (Bergman's, Golger's, etc.) observed in various animal groups.

A common pattern found in summary treatments of evolution in mid-century textbooks, such as Hickman (1955) or Villee (1954), was to present an outline of neoDarwinian processes as a mechanism of evolutionary change, often with equal space for other, clearly obsolete views. Thereupon any axiomatic use of the theory was abandoned, and it was merely assumed that natural selection always promoted what was in some way good. It made predator-vulnerable animals resemble their backgrounds; it made adults unselfishly devote themselves to reproduction so that their species might persist; it made reptiles supersede amphibians, and mammals supersede reptiles; perhaps it even assured the emergence of modern civilization (Huxley 1953, 1954). Explicit models of group selection, or any other kind of formal evolutionary models, were seldom proposed.

There were some commendable exceptions to the general intellectual laziness. The most important to me was the levels-of-selection discussion by Allee *et al.* (1948, p. 692). Only later did I learn about the serious work that had been done: Fisher's (1930) rigorously axiomatic discussion of natural selection (augmented in 1958 to include an explicit treatment of 'The Benefit of the Species'), Haldane's (1932, Appendix) model of the evolution of altruism, Sturtevant's (1938) use of group selection in relation to social insect evolution, Lack's (1954*a*, *b*) seminal field and theoretical work on the evolution of reproductive effort. In those days few people read such mathematical discussions as Haldane's and Fisher's. We got our theory in words from Dobzhansky, Mayr, and Simpson, who were deeply concerned with natural selection, but gave little attention to its levels of operation.

The neglect of the theoretical basis for group-benefit adaptations ended abruptly with Wynne-Edwards' (1962) arguments for selection at the level of local populations and against the adequacy of natural selection in its usual text-book formulation. Wynne-Edwards' thesis was that selection within animal populations could be expected to result in ever increasing effectiveness in garnering resources, usually food, and turning them into offspring. This would lead to overexploitation and the extinction of the excessively efficient and prolific exploiter.

Wynne-Edwards saw a close analogy between overexploitation of a fishery and what would be expected of natural selection within populations. The expected increase in efficiency and depletion of resources must be suppressed by a more powerful selection among populations. He argued that those that survive must be those that restrain their reproduction, so that crowding and destruction of resources are avoided. The constant weeding out of populations that adversely affect their food supplies would assure that the species as a whole acquires and retains birth control mechanisms capable of avoiding adverse effects on resources. Much of Wynne-Edwards' (1962) book was an attempt to show that animal social and reproductive behavior conformed to expectations of reproductive restraint. Animals almost always, according to Wynne-Edwards, reproduce at rates below their actual capabilities.

Much work by a legion of others since 1962 has shown the opposite, that animals and plants do their utmost to achieve maximal reproductive success (e.g. most of the references in Krebs and Davies (1984), and Clutton-Brock (1988). Animal fecundities are indeed less than they could be, but this is now generally attributed to trade-offs between numbers of offspring and resources allotted to each, to adaptive partitioning of resources to present and future reproductive opportunities, and to related tactics in optimized resource allocation.

4.6 An aside on beehives and haystacks

Having sketchily discussed selection at the material levels *trait-group* and *population*, I can now more easily consider groups that may seem to fall between trait groups and populations as usually conceived. Undoubtedly the most important kinds of trait groups arise from the inevitable associations of individuals in reproduction. Except for asexual or selfing species, a male and a female or two outcrossed hermaphrodites must get together to produce offspring. Offspring must be at least temporarily associated with at least the female parent (the yolking of an egg prior to fertilization is such an interactive association). Reproductive groupings may sometimes evolve into close-knit but temporary nuclear families, as exemplified by pairs of parents and dependent nestlings of many song birds. More rarely, nuclear families may evolve elaborate social systems, of which the colonial insects form an extreme example.

Some insect societies seem to refute my claim on pages 45–6 that trait groups do not reproduce. The departure of the old queen bee with a large retinue of attendants seems to be the binary fission of a highly organized trait group. The process is not what it seems, as is apparent in taking account of the male role in such colony fission. What looks like the mother colony that provides the migrants for the new one is headed by a new queen from that colony fertilized by one or more males from elsewhere. The original queen's daughters, which formed the original social system, will soon be replaced by her granddaughters, more closely related to one or more strange males than to the original daughters.

The fission of the colony is reproduction in the material domain, but there is no parallel reproduction in the codical. The colony is very much an interactor, and as expected, colony properties can be optimized (Nonacs and Dill 1990) in ways that justify the *superorganism* concept of Seeley (1989) and Wilson and Sober (1989). They are optimized for the propagation of their genes, not their genotypes or gene pools, and explanations for within-colony cooperation must be sought in effects on gene frequencies. Ratnieks' (1988) models of mutual policing by colony members is a good example. As long as there is genetic heterogeneity within social insect colonies, and frequent recombination between, their evolution can never be realistically depicted by a dendrogram. Selection among colonies is not selection among phylads.

Many large populations may consist of small and isolated groups that, unlike many social insect colonies, often do last more than one sexual generation. This is especially true of organisms with wide numerical fluctuations, lemmings being the classic example discussed by Bulmer (1988). At minimal abundance fertile pairs or pregnant females may be widely separated from each other. Distances may be such as to make finding unrelated mates, by the offspring of such individuals, hazardous and unlikely. At these numerical lows, lemmings or other small rodents may avoid leaving home (e.g. a local haystack), and close inbreeding may be the rule. Those local inbred groups that happen to maintain a consistent female majority will tend to increase faster than those with more males. When high numerical abundance over a large area is again achieved, there may be an overrepresentation of the local groups that had a female-biased sex ratio. If this cycle is often repeated, and if there is genetic variation in sex ratio, a consistent female bias may evolve.

This is the *haystack* model, originally of Maynard Smith (1964), recently discussed by Wilson (1987) and Taylor and Wilson (1988). The material events described above are represented in the codical domain by the multiple fission of a gene pool into many descendant gene pools that remain independent for a few (e.g. two or ten) generations. During this interval the isolated populations come to differ greatly in number (most no doubt reach zero), and their sex ratios may well contribute to this numerical variation. This can be recognized as clade selection for female bias.

While this is a valid interpretation, it should be realized that the clades are separate for only a short time on an evolutionary time scale. Modelers may find it more convenient to recognize a single population living in many haystacks for several cycles of abundance (Grafen 1985; Nunney, 1985*a,b*; Frank 1986; Bulmer 1988; Nunney and Luck 1988). Such models consider the fortunes of single genes over the entire cycle of population abundance and may provide the easiest way of calculating gene-frequency change over many generations. This is a feature of great value, but there are other values. I think it desirable, in thinking about organisms for which the haystack model is descriptive, to realize that selection in female-biased Mendelian populations favors males, and that it is only the selection among such groups that can favor the female bias.

4.7 Selection among more inclusive phylads

As argued in Chapter 3, the set of concrete organisms that collectively bears a single gene pool is an interactor on which selection can act. So is a group of organisms that collectively bears a monophyletic group of gene pools (phylads, of Table 3.1, p. 00). Such genetically related groups proliferate or die out, and it seems unlikely that alternative groups will be equally well equipped for long-term survival and cladogenesis. Any variation in group phenotypic fitnesses, with heritability of fitness differences, will provide the potential for natural selection, no matter what the taxonomic rank of the group. Group heritability is a measure of the persistence of character states through episodes of cladogenesis. There can be no question of the reality of phylad selection and resulting clade selection at all taxonomic ranks. The important issues relate to its exact nature and importance.

In this chapter I will illustrate the idea of clade selection by considering how it might be brought about by phylad differences body size, which Van Valen (1975) showed was subject to clade selection in the Mammalia (pp. 31–2). My example follows Harvey and Pagel's (1991, p. 42) suggestion that a larger mammal may fight back against a predator from which a smaller one would flee to a burrow. Suppose a woodland is inhabited by two kinds of rodent, one with an adult body size of about a kilogram and the other perhaps a quarter that big. Within each population there is normalizing selection for the current size. In the smaller species, for instance, smaller individuals may be less effective at competition with their fellows, and larger ones may find it difficult to maintain sufficiently commodious burrows, on which escape from predators often depends.

Unfortunately for this smaller species, a population of weasels gets established in the woodland, and it is capable of following the small rodents into their burrows. Those that do not flee into burrows are even more readily caught by the weasel, so that burrowing, while not very effective, is still the best available defense. The result, a few years after the weasel introduction, is the extinction of the smaller rodent. In the larger species the older juveniles and adults are able to fight well enough to discourage weasel attack, they can often defend their young against the weasels, and they persist after the smaller species is killed off. This to me is clade selection for gene pools that specify interactors with large body size.

To Vrba (1984) and Maynard Smith (1984), it is individual selection, because size is an individual character and because it is only size, rather than species membership, that causes the greater death rates of the smaller species. A 250 g juvenile of the larger species may be just as vulnerable as an individual of the same size in the smaller. All this is true but trivial. The important point is that there are two kinds of selection at work here (within and between populations), and it is necessary to distinguish them and convenient to give them different names (Fig. 4.1).

The two populations may be said to compete for the inattention of the predator. The larger form is the better competitor because of the greater strength and, probably, speed conferred by its larger size. The strength and speed of the smaller are of little avail, and it must rely on the rather ineffective tactic of retreating to a burrow. Clade selection can thereby favor a larger size, while individual selection remains strictly normalizing, as it was before the weasel appeared. Among individuals of the smaller species, it may still be true that those of the average size shown in Fig. 4.1 are favored. They would be contributing more genes to their gene pool, and this is favorable individual selection by definition (Fig. 2.2, p. 17). They may even survive weasel attack more often than slightly larger

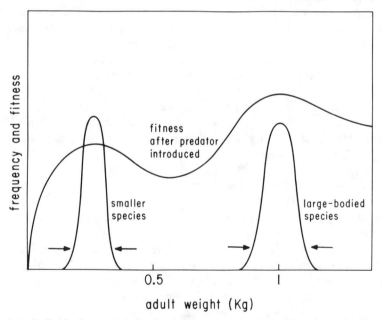

Fig. 4.1. Individual vs. phylad selection in a community with a large-bodied and a small-bodied species of rodent. The peak fitness attainable by the small bodied form may not be sufficient to enable it to avoid extinction. This is irrelevant to individual selection (shown by arrows), as long as slightly larger size reduces rather than augments fitness, as shown. The extirpation of the smaller rodent and survival of the larger would illustrate clade selection.

individuals, if the larger ones can squeeze into fewer burrows. The last individual of the smaller species to die (the ultimate winner of the individual-selection game) may even be a bit smaller than average.

Unlike Van Valen's (1975) facts on clade selection for small size in mammals (or Roughgarden and Pacala's (1989) on clade selection for large size in lizards), my fiction on clade selection tells why selection favored being big over being small. Even if one denies that Van Valen's data showed that size itself affected the evolutionary potential of mammalian taxa, it seems clear that large-bodied forms fared less well in clade selection and the small-bodied less well in individual selection. Selection at both levels took place in Van Valen's example, even if we do not know why, and it went in opposite directions at the two levels. Surely it would be unreasonable to call both processes individual selection, as implied by Vrba (1984) and Maynard Smith (1984).

As argued in Chapter 3, the sympatry in my hypothetical example is not necessary to the conclusion. Suppose my two rodents lived in different areas, but the weasel was introduced into both, with the same extinction of the small species and survival of the large. I am inclined to recognize that clade selection is operating even here, unlike Damuth (1985), who maintains that only sympatric *avatars*, populations in ecological competition, can be alternatives subject to selection. Allopatric forms may not be ecological competitors, for the inattention of a predator or anything else, but they compete for representation in the biota, the ultimate prize in clade selection.

I will venture a less hypothetical example from the mammalian biota. About 40 million years ago parts of North America were inhabited by enormous ungulates called titanotheres. I presume that they have no living descendants. Their place has been taken by bovids and cervids descended from Old World ancestors that were surely not in ecological competition with New World titanotheres. If this survival of one group and extinction of the other was the result of consistent differences in some kind of adaptive performance, it should be considered clade selection.

This last condition is the crucial one. The survival and extinction must result from collective differences in the adaptive performance of phylads so that their collections of gene pools differ in ability to survive as clades. This ability might be defined as the difference between the aptitude for reproduction (cladogenesis) and vulnerability to extinction, or perhaps the ratio. It would be detected and measured by comparison with expectations of an appropriate null model of random branching and extinction. It is unrealistic to expect that actual causes of any instance of clade survival and extinction can be known from the fossil record. The best that can be done is what Van Valen (1975) did, and look at a large sample of survivals and extinctions for empirical rules, for instance a relationship to body size. If such efforts are frequently successful, it will prove that clade selection has been important in determining the kind of biota that now inhabits the Earth. If they are usually unsuccessful, it will indicate the macroevolutionary unimportance of clade selection.

One final point on the meaning of clade selection. As I have used the term, it is a concept a bit broader than competitive exclusion or ecological succession, because it does not require that selection act among ecologically competing entities. It differs in practice in its emphasis on the geological rather than the ecological time scale, and perhaps on a broader spatial perspective. Also, studies of ecological succession do not normally allow for major changes in the ecological niches of the populations that succeed each other, but this would be routine in considering clade selection. The main difference may be simply in the attitude and purposes of users of the concepts. Dawkins' (1982a) *Necker cube* analogy may be apt here.

4.8 Clade selection and punctuated equilibria

Clade selection is a topic of broad scope and importance, but past discussions have been limited mainly to two special contexts, the species level of taxonomic distinction, and the punctuated equilibria theory of macroevolution. Concentration specifically on *species selection* (Stanley 1979; Vrba 1984, 1989) is one of several serious abuses of the species concept (Chapter 8). As most readers already know, the idea of punctuated equilibria of Eldredge and Gould (1972; S. J. Gould 1982b) is a complex and controversial topic, and I prefer to neglect it except for the brief treatment in Chapter 8 and the following discussion of its relevance to clade selection.

It is undeniable that rates of evolution are immensely variable, a topic treated in exemplary fashion by Simpson (1944). It is a matter of observation that characters of some species of higher animals and plants can evolve in domestication at rates of several per cent per generation and in nature at several per cent per century. Even at one per cent per century a mean value could double in 7000 years. Any such change would appear in a typical fossil record as a large and sudden saltation. It is understandable that the fossil record will often fail to document the origin of what seem to be quite distinct new forms. If most evolutionary rates are far slower, so slow that a doubling might take the entire Pleistocene, even well documented phylogenies may approximate the rectangular patterns pointed out by advocates of punctuated equilibria (Fig. 2.1C).

Another factor that I would assume to be true is that rapid evolution is often caused by rapid environmental change. This rapid evolution in closely related clades need not always be in the same direction. For at least some characters the close relatives will diverge rapidly in times of rapid environmental change. Such times will also be periods of stress that may cause frequent extinction. It is entirely to be expected, from what we think we know about evolution, that rapid (even seemingly instantaneous) character change, rapid cladogenesis, and frequent extinction would be closely associated in the history of life. In this sense it is inevitable that important evolutionary changes will be associated with cladogenesis. Moreover, unless extinctions are completely independent of the characters evolved by the different clades, clade selection, especially at the lower levels such as species and subspecies, will be especially strong at times of rapid change (of course 'especially strong' may still be extremely weak from an experimentalist's perspective, as discussed in Chapter 3). There is nothing in the envisioned process that makes rapid evolution and major macroevolutionary change dependent on cladogenesis.

4.9 Genealogically mixed interactors

Members of genetically separate populations can sometimes form groups important as interactors for genes, gene pools, or clades. Such interactors can arise from mutualisms that become obligate. Frequently claimed examples are the fungus–alga combination in a lichen, the termite and its gut biota, the polyp–zooxanthella association in a reef coral. How mutualisms between separate species become specialized and obligate, and how they get started in the first place are complex questions, discussed by Colwell (1986), Wilson and Knollenberg (1987), Boucher (1985; and other papers in the same volume). An important insight was recently provided by Ewald (1988), who argued that parasites and commensals will tend to evolve into mutualists to the extent that they come to depend on the host's successful reproduction for their own transmission. This is an aspect of genetics being a limiting case of epidemiology pointed out above (p. 15). Williams and Nesse (1991) reviewed other work on evolution along the parasitism-to-mutualism continuum.

A simple test may make it possible to decide whether an association of originally separate populations is not only an interactor, but one that has its own combined codex. If two associates tend to speciate together, they can be assumed to have joined forces to produce a common interactor that is now essential to their originally separate codices. Cockroaches and their endosymbionts apparently speciate together and a cladogram for one closely corresponds to a cladogram for another (Wren *et al.* 1989). This is apparently not true of most other multi-species associations (Stone and Hawksworth 1987). It would appear that recognition of lichens as a taxonomic category is invalid. The fungi and algae have their own separate phylogenies (Lawrey 1984; Amadjian 1990), as do corals and zooxanthellae (Rowan and Powers 1991), and angiosperms and their root symbionts (Nap and Bisseling 1990).

An ecological community of more than minimal complexity will always provide many examples of mutualisms and of unilateral dependencies of one population on one or more others. Theoretical and experimental models (Goodnight 1990; Setala and Huhta 1991) indicate the possibility of community-level selection in macroevolution. Whether it actually operates in nature, strongly enough to make a community an entity for which the concept of functional design would be applicable, is another question. Even with a high local incidence of mutualisms, competition between mutualistic coalitions may select for largely negative interactions and thereby compromise the fitness of the community as a whole (Dodds 1988). The overwhelming prevalence of the negative interactions of parasitic and predatory exploitation indicates that communities have not been formed by selection for any sort of efficient and harmonious operation.

The reason must be that communities lack the necessary high rates of reproduction and replacement and especially the high level of heritability required for effective selection. They change their makeup so rapidly that selection among communities must be overwhelmed by endogenous change. Mammalian communities merely in the late Quaternary, according to Graham (1986) 'have been massively and repeatedly reshuffled.' Brown (1987) says essentially the same for desert rodents, and Davis (1989) for plants and other groups. Other evidence of rapid flux of community composition in recent millennia is reviewed in Chapter 9. The venerable idea that communities are organized by competition and coevolutionary niche partitioning has received little support from recent studies (Underwood 1986; Gee and Giller 1987; Simberloff and Connor 1981; Yodzis 1981; Drake 1991).

5

Optimization and related concepts

I argue in Chapter 6 that organisms are never optimally designed. Designs of organs, developmental programs, etc. are legacies from the past and natural selection can affect them in only two ways. It can adjust the numbers of mutually exclusive designs until they reach frequency-dependent equilibria, often with only one design that excludes alternatives. It can also optimize a design's parameters so as to maximize the fitness attainable with that design under current conditions. This is what is usually meant by optimization in biology. An analogy might be the common wooden-handled, steel-bladed tool design. With different parameter values it could be a knife, a screw driver, or many other kinds of tool—*many*, but not *all*. The fixed-blade constraint would rule out turning it into a drill with meshing gears. The wood-and-steel constraint would rule out use as a hand lens.

5.1 Frequency-dependent selection

Maynard Smith (1974, 1982) is both pioneer and current doyen of ESS (Evolutionarily Stable Strategy) theory (see also Parker 1984; Hines 1987), which now provides the standard method of analysis for *frequency-dependent* equilibria. This theory would have served Darwin well in coping with special difficulties he perceived (1859, Chapter 6) in 'diversified habits in the same species.'

Prevalence of maleness and femaleness as mutually exclusive designs is the most easily appreciated example of an ESS resulting from frequency-dependent selection. Any frequenter of a singles bar knows that membership in the minority sex confers an advantage in the games played in such settings. The same principle applies to a natural population. If there were consistently more females than males, the males would, on average, reproduce more successfully. Selection would favor any tendency to be male or to produce sons rather than daughters. The result would be increasing male frequency and decreasing male advantage. At the stable equilibrium (ESS) the sex ratio would assure that neither sex has an advantage over the other. Note that when this frequency-dependent selection has reached the ESS there is no adaptation to an environmental condition. Selection destroys the condition, such as excess of one sex among conspecific neighbors.

A less hackneyed example is found in Batesian mimicry (Turner 1978, 1984). A noxious, inedible butterfly (the model) evolves a conspicuous coloration that deters potential predators. It may then have this coloration closely imitated by an edible compatriot. This *mimic* enjoys the benefits of potential predators' having learned to avoid its warning coloration. Unfortunately for both mimic and model, the predators may learn from the mimic that the special coloration means edibility. The greater the prevailing mimic/model ratio, the greater will be the tendency of predators to attack both. This effect can be reduced if the mimic can imitate more than one inedible model, and this possibility has been exploited by a number of mimic species. If there are two color phases, resembling the warning colors of two different models, the equilibrium for the mimic will be that ratio of the two phases at which neither does any better than the other. This ratio will be set by the relative numbers and unpalatabilities of the models and by other factors. Many different kinds of natural selection are convincingly shown by the phenomena of mimicry (Turner 1978).

ESS reasoning is applicable whenever the adaptive values of attributes of one individual are conditional on those of others, even when selection is not entirely frequency-dependent. If there is density dependence along with frequency dependence, evolution can select for optimum phenotypes rather than stable equilibria. An example is selection for egg number when two parasitoids may lay eggs in one host. If survival rates are a linearly decreasing function of egg number, the optimum number of eggs for each female will be two thirds of what it would be if she were the sole user of the host (E. L. Charnov, personal communication).

This is density-dependent, not frequency-dependent selection. A high population density (two rather than one egg layer per host) results in resource depression, and this has an understandable effect on optimum clutch size. At equilibrium, any deviation from the optimum results in lower fitness, but with frequency dependence, deviations from equilibria are adaptively neutral. For instance in a large population at its equilibrium sex ratio, it makes no difference whether a parent has all sons, all daughters, or any mixture of both. An optimum would appear as soon as there is the slightest deviation from the ESS. Even a minuscule excess of males, for example, makes an exclusively female progeny the optimum parental strategy. Equal numbers of sons and daughters would never be optimal in a large and fluid population.

Given my special concept of adaptation as conformity to *a priori* design specification (Chapter 4), the following tenets of frequency-dependent selection theory should be emphasized for their relevance to later discussions.

(1) The theory in no way accounts for the evolutionary origins of the alternatives it considers. It accepts them as historical legacies and deals with their relative numbers.

(2) It is irrelevant to the ontogeny of the strategies considered. It may tell us why manhood and womanhood or other sexual phenotypes have certain frequencies, but it leaves untouched the question of whether sex is environmentally or genetically determined, or even whether there are men and women, rather than a hermaphrodite population in which each individual partakes of both manhood and womanhood.

(3) More than one strategy or ratio of strategies may be a stable equilibrium. The one established by selection may depend on the starting ratio, another kind of legacy from the past.

(4) Game theory was designed to enable engineers and economists to arrive at the one *best* solution to a problem. Evolution seeks only the most easily attainable *stable* solution, which need not be best, nor even moderately good, for either population or individual (Hammerstein and Parker 1987). Evolution, in reaching a frequency-dependent equilibrium, either fixes one strategy because it wins over all available alternatives, or establishes a stalemate between competing strategies.

(5) An ESS payoff matrix by itself says nothing about evolutionary changes in descriptive details of strategies considered, such as maleness and femaleness. It can tell us about relative numbers of men and women, but nothing about specific male and female adaptations.

(6) As noted above, a frequency-dependent equilibrium is not an optimum, for either an individual or its population, nor is it the solution to any kind of problem faced by an organism. It simply abolishes the problem, and when the population is at its ESS there is no longer any selection on the frequencies of alternatives (Lloyd 1988).

(7) As noted by Maynard Smith (1982), the study of adaptation must frequently begin with an ESS analysis for an understanding of the relative frequencies of distinct forms (or exclusive prevalence of a single form) within a population (males and females, sexual and asexual reproduction, etc.). The next step is to apply the principle of parameter optimization, discussed at length below.

The difference between a frequency-dependent equilibrium and an optimum is nicely illustrated by the game of *prisoners' dilemma* (Fig. 5.1). C (cooperate) could mean willingness to give two units of aid to another individual, at one unit of cost to oneself. Two Cs engage in mutual aid that gives each a net gain of one unit per encounter. D (defect) could mean accept the aid offered, but do not reciprocate. In each encounter with a C, D gains from the aid offered but is free of costs (net gain two units). The C in this case pays the cost but gets no benefit (net score minus one). Ds encountering Ds have neither costs nor benefits (net score zero).

Now imagine three populations of cooperators or defectors. In the first, every individual cooperates at every encounter (the *Dr Pangloss* solution, favored by clade selection). It gets the highest average fitness, one unit of gain per encounter. In the second population, strategies are chosen entirely at random, C and D each having probabilities of 0.5. The four payoffs are all equally likely, and the long-term average payoff for every player is 0.5, the average of the four possibilities. The third population is at the ESS, as expected of the selection of alternatives within the population. Every individual plays D every time, and the payoff for every individual is zero per encounter.

Natural selection produces a population of defectors because it favors the higher-scoring strategy in every contest. With the payoff matrix illustrated (Fig. 5.1), an individual always does better by playing D, no matter what the opponent plays. D always wins over C, until the population consists entirely of Ds. Natural selection not only fails to maximize fitness, it minimizes it, making it considerably lower than would be accomplished by a purely random choice of strategy, and, in fact, lower than it would be with any other proportion of Cs and Ds.

The prisoners' dilemma has many biological applications, of which I

	C	D
C	1	−1
D	2	0

Fig. 5.1. Pay-off matrix for a game of Prisoners' Dilemma. Pay-offs, per convention, are for the player to the left. Natural selection consistently favors D (defect) over C (cooperate) and thereby minimizes average fitness, as explained in the text.

will discuss one. Small fish in a pond may have an optimal school size, determined by a trade-off between anti-predator benefits from gregariousness and competition for food among members. A lone individual is especially vulnerable to predation, but if it joins a group it will have others to compete with it for the attention of a predator. Strategy C for this fish might be 'join the first available group of below optimal size.' Strategy D might be 'join the first available group.' If the extra time required to find a small group imposes a predation cost greater than the feeding cost of membership in an oversize school, D will be fitter than C in any pairwise comparison.

If a C is already in an oversize school, it might leave to find a smaller one, and thereby expose itself to increased predation. A D in the same school would stay behind and benefit from the group now being smaller by one but would not pay the predation cost. A formally similar argument may be made for gregarious predators such as lions and wolves, if single individuals are at a great disadvantage in hunting. It is not surprising that the group sizes of gregarious animals are often larger than their ecological optima (Sibley 1983; Mittlebach 1984; Pulliam and Caraco 1984).

5.2 Parameter optimization

My distinction between designs and their parameters conforms to Gross' (1987) insistence that the term *strategy* be used only for characters or character complexes maintained by frequency-dependent selection. Each strategy is composed of *tactics*, and any quantitative aspect of a tactic, in Gross' usage, could correspond to one of my strategy parameters subject to optimization. Numbers of men and women in any human population are established by frequency-dependent selection. Parameters of maleness and femaleness, such as degree of development of secondary sex characters, would be selected to optimize trade-offs among mating advantages, nutritive and viability costs, and related factors. The venerable concept of *normalizing selection* (Avers 1989) would be closely related, but parameter optimization is perhaps more explicitly tied to quantitative characters and is more clearly predictive and vulnerable to data.

Monomorphy can be established either by frequency-dependent selection for a form that can resist displacement by all available alternatives, or by optimization. Optima would include not only values of maximum physiological effectiveness, but also those of maximum social acceptability. For instance, having a five-digit hand with an opposable thumb is no doubt a human optimum, rather than a frequency-dependent equilibrium. This need not depend on anything mechanically superior about the five-digit design. All that is needed is that individuals with other than five-digit hands have a mating disadvantage. An equally well engineered hand of four or six digits might work better. A properly engineered octopus-like tentacle with suction cups might work better still. These are irrelevant fantasies, because evolution works only with available alternatives, not hypothetical ideals.

There is no selection for having five fingers until someone with a different number, e.g. six, comes along. Human developmental genetics is such that this deviant individual would probably be mechanically defective in a number of ways, besides having an extra finger and consequent social stigma. The five-digit phenotype would be functionally superior and the alternative eliminated, thereby re-establishing monomorphy. The five-digit hand is a local optimum, functionally superior to any slightly different condition.

But suppose a miraculous mutation occurred that always caused its carrier to have six-digit hands and that such hands were very slightly superior to the ancestral form. My guess is that five digits would now be maintained as a frequency-dependent equilibrium. The miraculous gene would be eliminated because of the social stigmata and mating handicaps for six-fingered individuals in otherwise five-fingered populations. If ecological contingencies and genetic drift isolated a small population in which the six-fingered individuals were in the majority, it would be those with the ancestral hand design that would be discriminated against. This would be an example of the importance of initial conditions for frequency-dependent selection (point (3) in the previous section).

Natural selection acting on variation in a quantitative character (parameter of a design or strategy) can often be expected to fix the mean value near the functional optimum and to minimize the variation about that mean (qualifications in relation to genetic variation and other factors are considered in Chapter 6). Use of this idea is often called *optimality* modeling. I prefer the term *optimization*, which implies a corrective tendency, rather than a state attained. Likewise I think that calling a character *optimized* is always more realistic than calling it *optimal*.

If natural selection works with better vs. worse for design parameters of the human hand, it might be conceived as a force tending to optimize the relative lengths of each digit and each segment of each digit, the exact positioning of muscle attachments, and many other measurable features. If this is really true, some idealized investigations would yield predictable results. Suppose we could go back some tens of thousands of years and record two items of information about every member of the human species for many generations: lifetime reproductive success, and the length of the second phalanx of the third finger. When we plot our data on a scattergram, what form do we expect this scatter to take? The honest and straightforward prediction is that there would be little variability in the part measured, but that accurate measurements and elaborate statistics would show that those with close to mean values leave the greatest average number of descendants. If appreciable directional selection was taking place, there might be a slight difference between optimum and mean.

This may be an honest use of the concept of optimization, as outlined above, but it is also rather naive. It ignores much of what we believe to be true of human natural history in the Stone Age. We believe, for instance, that individuals always need to develop and grow. The length of each digital phalanx changes in development from microscopic at first appearance to whatever it reaches in its final, possibly arthritic form. Measuring each individual at the same age might correct for developmental change but would not confront some more serious issues that arise from human natural history and expected effects of natural selection. What do we really expect to be optimized? Mean absolute length of the bone at some age? Its length relative to the rest of the finger? Relative to the rest of the body? Is there just one such optimum for each sex, age, habitat, occupation? How many different optima for different circumstances can selection maintain for one parameter?

Surprisingly many, even if we confine ourselves to everyday observations and ignore the examples discovered by an enormous number of physiological studies. Instead of suggesting a bone in the hand, I might have suggested that we measure the diameter of a pupil. It might have turned out that human races found in, say, the forested and often cloudy regions of northern Europe evolved large pupils, of a size appropriate for average conditions of illumination in that habitat. By contrast, populations in the Sahara might have smaller pupils, better for the average conditions there. In fact, all human races (like all mammals) adapt to differences in illumination by moment-to-moment alterations of pupil diameter, from about 1 to 10 mm, which gives about a hundred-fold difference in light-gathering capacity. Selection maintains a great range of optima for a great range of conditions, from bright sunshine to nocturnal darkness.

I belabor these well known facts to make the point that one does not treat the optimization principle as a formula to be applied blindly to any arbitrarily selected attribute of an organism. It is normally brought in as a way of expanding our understanding from an often considerable base of prior knowledge, as was pointed out by Krebs and Dawkins (1984) and Harvey and Pagel (1991, p. 9). The same point was made by Symons (1987), who noted that human arterial blood is red and then asked rhetorically why it is that 'Evolutionists do not offer adaptive explanations for redness, physiologists do not study how redness works in the body, and developmentalists do not consider the ontogeny of redness to be an interesting question.' The answer he suggests is that 'we intuitively perceive the redness of arterial blood to be an *arbitrary* trait; by picking it we fail to carve nature at a joint.' He contrasts the redness of blood with the action of the human larynx in preventing the swallowing of food from clogging the respiratory passage. To Symons, the recognition of the swallowing mechanism does carve nature at a joint and provides a biologist with a phenomenon worth study. I presume that Symons would agree that we would expect various features of the swallowing mechanism to be optimized, for instance the timing of nerve impulses to each laryngeal muscle. This optimization allows us to get by most of the time with a grossly maladaptive legacy from protochordate ancestors (Chapter 6).

I offer Fig. 5.2 as a heuristic for the optimization process. Milkman (1982) offered a basically similar but more complex and more widely valid model. The trait considered could be any of a wide variety of plant or animal measurements, such as date of flowering in some angiosperm population. I assume a single optimum flowering time, rather than two or more maintained as a mixed-strategy ESS or from different optima reached facultatively by individuals in different circumstances.

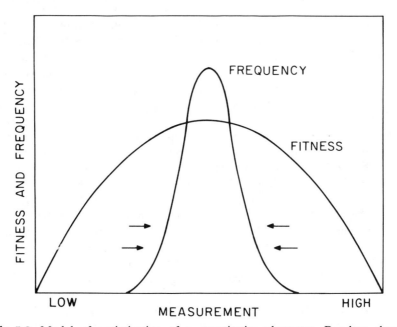

Fig. 5.2. Model of optimization of a quantitative character. Random change (mutation and recombination) tends to increase variability. Natural selection, indicated by the arrows, tends to minimize variability and keep the mean near the fitness maximum. A more realistic analysis, which recognizes that differences can be adaptive, is offered in Chapter 7.

The optimum date would be that which achieves the ideal compromise between various environmental hazards and opportunities: number and quality of possible pollinators, effects of competitors for the attention of pollinators, likelihood of favorable physical conditions for seed maturation, etc. Reasons why the model is invalid for many applications will be discussed later in this and in the next two chapters. Here I will defend its general validity for applications that make sense in relation to what we already know about an organism. This defense argues two points, that heritable variation is generally permissive of optimization by selection, and that such optimization has happened for most of the characters for which it is expected.

The adequacy of genetic variation to sustain the optimization process is the subject of a detailed recent review edited by Loeschcke (1987). The authors deal with the phenomena of genetic constraint (discussed further in Chapter 6), which might be defined as any restriction on a population's responsiveness to selection. One universal constraint is simply the limit on additive genetic variation present at a given moment. Even quite strong artificial selection can seldom alter a character by more than a few per cent per generation (Yoo 1980). This is an unimpressive constraint. A few per cent in a thousand generations would be extraordinarily rapid evolution in the fossil record. *Genomic largesse* (Turner 1977) might be more appropriate here than *constraint*. Rate of origin of new variation can also be considered more than ample (Lande 1975; Lynch 1988). The logical difficulty with additive genetic variation and responsiveness to selection is that there is too much, not too little (Chapter 9).

More interesting constraints result from pleiotropy. Selection for one character may change a second that is seemingly unrelated. If the change in the second character is maladaptive it will limit or slow what can be accomplished by selection on the first. Loeschke's studies generally support the idea that pleiotropy and other constraints may markedly curtail the rate of evolution resulting from natural selection, but have little influence on the direction or ultimate outcome. Use of the optimization idea seldom depends on how fast or by what trajectory a character may approach its optimum under natural selection.

Charnov (1989) has analyzed the logic of quantitative-genetic and optimum-phenotype models and concluded that the optimization concept gives a deeper understanding of why organisms have the phenotypes that we observe under the usual process of normalizing selection. Mayr (1983) has pointed out that the main justification for the *adaptationist program* is that 'The adaptationist question', what is its function?, has guided every advance in physiology for centuries. Harvey's detailed demonstration of the mechanics of blood circulation in 1628 is an early example.

Justification for the adaptationist program need not be sought in the success of any recently formulated concept in behavioral ecology, such as optimal foraging, but this and other recent uses of the optimization idea do add effective weapons to the adaptationist's arsenal. A generalized adaptation concept merely predicts that animals may forage productively and, sometimes, adequately for their own needs. An optimal foraging model may predict such observations as just how heavy a load of food a bird will carry to its nestlings or the amount of time a dungfly will spend in copula (the *marginal value theorem* of Charnov (1976; see also Krebs and Davies 1987, Chapter 3; Horan 1989; Cuthill *et al.* 1990). Magnitude of parental devotion is apparently a design parameter subject to optimization in fish (Pressley 1981; Sargent 1988), birds (Thornhill 1989; Pugesek 1990; Wiklund 1990) and mammals (Smith 1987). A striking example of the value of optimization modeling is Belovsky's (1984) successful prediction of the range of trunk diameters of trees harvested by beavers as a function of distance from the pond. Checking on the validity of this kind of prediction is much more conducive to the advance of science than is the observation of some animals' getting enough to eat, noteworthy though that really is.

Related arguments for the adaptationist program, and by implication the optimization concept, have a long history (Muller 1948; Tinbergen 1965; Cain 1964). For more personal support of the idea that design parameters are optimized, I invite further consideration of my original example, the human hand. Imagine various quantitative alterations: in the positions of various ligaments or insertions of tendons from distant muscles in the forearm, any rerouting of blood vessels or nerves, alterations of tissue distributions or of the sizes, shapes, or positions of the nails. If this exercise fails to convince, I suggest a reading of the brief selections in the Appendix, and recent works on the interpretation of variation in populations (Chapter 7).

5.3 Character values and fitness values

Figure 5.2 and the foregoing discussion conform to some common suppositions: (1) some specific value of a character results in maximum fitness; (2) deviations from this optimum depress fitness at an accelerating rate; (3) equal deviations, whether positive or negative, result in equal decrements to fitness; (4) phenotypic effects of environmental variables are moderate, perhaps with the standard deviation equal to a small fraction of the mean; (5) high and low values of a character are achieved with about the same expenditure of resources; (6) the character is subject to polygenic influences, at least a trace of which is additive; and (7) selection, having operated under these conditions, will have fixed the population mean near the peak of the fitness distribution and restrict variability to a narrow range. The seventh proposition (parameter optimization) requires the others except for number two.

This second item, with twice the deviation causing more than twice the loss of fitness, is not necessary but has intuitive appeal. It is commonly modeled as fitness decrement proportional to the square of the deviation from optimum (Crow and Denniston 1981; Crow 1987; Rose *et al.* 1987). The first condition would be violated if there were two or more optima, often mapped on *adaptive landscapes* (Wright 1932; Provine 1986). More information would then be needed to decide what selection ought to bring about. It might, for example, produce a population with both optimum phenotypes and few or no individuals near the middle of the intermediate range. If the heights of the optima depended on frequency or density their relative numbers should then conform to an ESS. If the optima were close relative to variability, selection might accomplish nothing better than a variable population with a mean somewhere near both peaks (Via 1987).

I will delay discussing the problem of cost of production (pp. 69–71) and deal here with violations of the third and fourth propositions, on symmetry of fitness distributions and reliability of developmental programming. It may often be that one unit of excess will depress fitness more than one unit of deficiency, or vice versa. Extreme examples have been referred to as the *cliff edge effect* (Boyce and Perrins 1987). For a technological analogy, suppose you were required to design an airplane to land on a 1-km landing strip. The far end of the strip is bounded by a stone wall. There is no harm in bringing the plane to a stop in 900 m, or 800, but there are costs to shortening the required distance (more expensive breaks, reduced payloads, etc.). The engineering ideal is a plane that can always stop at about 999 m, a precision unattainable under any conditions, and, even if it were, conditions often vary (e.g. headwinds vs. tailwinds).

I presume that decisions on the plane's design would be influenced by the relative costs of various breaking capabilities and of collisions at various speeds with the stone wall. The resulting plane might then have an average stopping distance of 800 m with a standard deviation of surely less than 100 (Fig. 5.3). The moral of the story is that a series of measurements of the stopping distance of our plane, or of several from the same fleet, would show a mean value far below the one of maximum fitness, 999 m. Stearns (1976, Figs 2 and 3) gives reasons for believing that clutch sizes in birds may have the same kind of fitness asymmetry.

Yoshimura and Shields (1987) propose a convincing model applicable to asymmetrical fitness distributions and apply it to pollen dispersal. Maximum pollination success in their example was achieved at distances

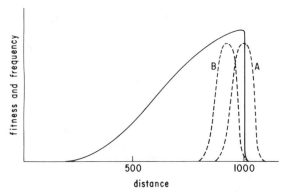

Fig. 5.3. Performance of alternative designs of airplanes required to use a 1 km runway with a stone wall at the far end (1000 m position) and with increased costs of manufacture for increased breaking capability. The continuous curve shows fitness as a function of stopping distance. Dotted curves show stopping frequency distributions for two designs. Design A has a mean stopping distance at the point of maximum fitness. Design B has its mean stopping distance well below the optimum, but it has a much greater mean fitness, measured as the mean product of the fitness and frequency curves.

of 3–10 m. Pollen grains dispersed less than 1 m were likely to reach stigmas already pollinated by the same plant, or perhaps they resulted in inbreeding depression. These could cause minor losses of fitness. Pollen dispersed more than 10 m was likely to reach habitats without any receptive stigmas, and this would be a major waste and loss of fitness. Dispersal of pollen is such a chancy process that closely similar grains released within a short time from a single source will often differ widely in distance traveled. Those with a mean in the optimum range (3–10 m) would scatter more often to unproductive distances than those with a lower mean dispersal. It is not surprising that the observed mean dispersal distance was below that of maximum fitness.

Although I think the model proposed by Yoshimura and Shields is intuitively most appealing with an asymmetrical fitness distribution, it is applicable to phenotypic probability distributions in general. It would be useful for symmetrical distributions if there were a tradeoff between mean and variance. A genotype with a mean phenotypic value at peak fitness might have a lower mean fitness than one with a less than ideal mean but lower variability, so that it would produce really low-fitness individuals less often. Barkan (1990) interprets chickadee foraging parameters as designed more to avoid dangerously low feeding rates than to maximize mean rates. Biologists testing optimization models ought always to bear in mind that the genotype with the greatest mean fitness will be the one with the highest frequency-weighted mean of phenotypes produced. This need not be one with its phenotypic mean at maximum fitness.

The likelihood of an asymmetrical fitness distribution of female choice criteria may resolve one of the major theoretical questions about sexual selection, a matter discussed in Chapter 8. Here I will apply it to one other theoretical challenge, the prevalence of play in the animal kingdom. Many biologists have attempted to explain why play should have evolved, and most have sought adaptive explanations, such as play providing needed practice or information for the development of adult behavior. Fagen (1981) gives a thorough review of the ideas and phenomena. More recent work is summarized by Barber (1991). I find the adaptive explanations unconvincing, because they seem to rely on arbitrary postulation of needs that are best satisfied by play. Barber (1991), for example, assumes that mammals have excess energies that have to be dissipated and that the development of the sympathetic nervous system would be frustrated without play. I prefer to regard play as an incidental consequence of an economy of developmental programming.

If a capability is needed at a certain stage of life and not before, a simple optimization model would predict it to make its appearance instantaneously in complete form at that stage. If the need and importance arise gradually, a gradual development of the capability would be expected. By this sort of reasoning play must be adaptive. A kitten's chasing a blowing leaf may never be immediately effective in procuring food, but so universal a kind of kitten behavior is not to be expected unless it is adaptive in some way.

Why not apply the same sort of reasoning to anatomical features? The kitten has nipples and spermatogonia or other reproductive structures. Must their early presence, long before they assume their usual adult roles, have some adaptive reason? I suspect that most biologists would agree that they are there, not for any contribution to fitness in the kitten, nor for any kind of present capability that can be used to enhance future fitness, but because they are imperfect precursors gradually developing for their future usefulness.

It may be theoretically possible for most of the development of the cat reproductive system to be delayed until, if not the moment before, perhaps the week before its first effective use. I suspect that this would be a more costly developmental process, if not of materials, at least of genetic resources (Waddington and Lewontin 1968; West-Eberhard 1986, pp. 1389–90). Precisely timed developmental processes of extreme rapidity would no doubt require closer allele-frequency adjustments at more genetic loci than a leisurely process with less critical timing. The various components of a cat's reproductive system have to be ready and functioning at a particular time in its life history. If a component attains adequacy a day too soon the consequences for fitness are negligible. If it becomes adequate a day too late the result may be disastrous, like my plane's collision with the stone wall (pp. 66–7). I would expect the cat's development to be programmed to assure that each needed component is ready when needed. If this results in most components being ready shortly before they are needed, and partly developed long before, the cost may be less than those required for more rapid and precisely timed development.

Reasoning applicable to juvenile reproductive structures, or to the very existence of prereproductive life-history stages, should be as applicable to juvenile reproductive behavior. It might be theoretically possible to evolve a tomcat that would pursue a life of perfect chastity and brotherhood until a precise moment of sexual maturity, when its behavior would suddenly change to what we expect of adult tomcats. I also assume that this program of behavior development would be more costly than having courtship and mating behavior and sexual aggression develop more gradually during juvenile life. Play exists because allowing behavior patterns to appear too soon is a low-cost imperfection in a cost-saving program for producing adaptive behaviors soon enough. The morphogenetic analogy is my own, but the explanation for play is one of several discussed by Ewer (1968).

5.4 Strategies, tactics, and winnings

My final qualification of the optimization principle relates to the expense of character attainment and the distinction between characters and winnings. If development of a character is a significant drain on resources (violation of condition 5, p. 65), the mean value shown by a population may be less than that of maximum fitness, perhaps far less. A frequently valid example is body size. In many organisms, reproductive success is strongly correlated with size. This is generally true in a simple way for female animals and most plants. The larger they are the more eggs they can produce, and the more eggs the more zygotes. If there is no postzygotic investment by a female animal, her zygote production may realistically measure reproductive success. The same can be said of seed production in most female plants (Solbrig and Solbrig 1984).

The exact effect of size on fecundity is conceptually complex and may be difficult to measure. Productivity of eggs must be a power function of size, like any other kind of productivity (Taylor and Williams 1984), and a doubling of mass is likely to have a less than doubling effect on fecundity. Measurements of fecundity will show effects of size only if certain other factors, for instance priorities for resource allocation, are independent of size. A large fish may be more fecund than a small one, not only because it is more productive, but also because it puts more of its productivity into eggs and less into somatic growth.

Larger size can also mean larger production of sperm or pollen. This can be important with mating systems in which male success depends largely on sperm or pollen competition. Yet even in the more extreme examples, male success will usually be less size dependent than female success. This is because a male is usually his own worst enemy in sperm competition. When a sperm enters an egg, most of the nearby sperm that are excluded may be from the male that provided the successful sperm. There would seldom be comparable egg competition. One egg's capturing a sperm would not normally prevent another egg from being fertilized. Thus the fecundity advantage of large size is usually greater for females than males, and this explains the general tendency in the animal kingdom for females to be larger than males, sometimes extremely so (the dwarf-male phenomenon of Ghiselin (1974a)). Where sperm competition is not the only way males compete with each other, the size-advantage relation to sex may be reduced or reversed. For instance in polygynous mating systems a bit of extra size may enable one male to sequester one or several mates while a slightly smaller one remains celibate. Polygynous mating systems, such as are often found among mammals and in a minority of fishes and a few other groups, can cause the evolution of fully mature males larger than females.

Size is perhaps the most generally valid example of an expensive character. Far more material must be devoted to the production of an animal of 4 kg than one of 2 kg. For most organisms it would be entirely unrealistic to expect a population's mean adult body size to be close to that of maximum viability or reproductive performance. What would that size be for a halibut or a white pine? Theoretically, continued growth would produce a halibut or pine so big that its species' normal maintenance mechanisms are taxed to the point of balancing any fecundity advantage to still larger size. This optimum is no doubt seldom reached by real halibut or pine trees, and it may be that most adults have far smaller sizes than those that would maximize adaptive performance.

When this is true the measurement in question is not of a character optimized for its role in maximizing fitness, but rather an index of current fitness attainment. A size ranking among similar-age halibut or white pines must be very nearly a ranking according to phenotypic fitness. The same may be true of many other characters closely associated with viability or reproductive performance: a lemming's fat reserves at the start of hibernation, antler size in deer, numbers of flowers on a plant, numbers of ramets in a clone of strawberries, and perhaps clutch size in birds (Stearns 1976; Price and Liou 1989). Such characters are neither tactics nor strategies but rather measures of current winnings in reproductive competition. Heritabilities of such characters closely predictive of fitness need not be zero by the usual measures (Gustafsson 1986; Mousseau 1987; Gibbs 1988; Cooke *et al.* 1990) but they must be low. Otherwise they would increase rapidly with the passage of generations.

6

Historicity and constraint

This book is about natural selection, with all discussion constrained by acceptance of the two other doctrinal bases of modern biology, mechanism and historicity. In accepting mechanism I assume that all the laws of chemistry and physics are obeyed by living organisms, in their individual activities, in the workings of their minutest parts, and in their ecological interactions. This chapter will deal with the more generally unappreciated doctrine of historicity and with related evolutionary constraints (phylogenetic, developmental, genetic) on the action of natural selection.

6.1 Organism as historical document

As noted in Chapter 1, every organism shows features that are functionally arbitrary or even maladaptive. I illustrated this idea with two examples from mammalian necks, the number of cervical vertebrae and the crossing of digestive and respiratory systems. Here I will present two others, a classic anatomical example and another less widely appreciated.

My chosen classic is the vertebrate eye. It was used by Paley (Appendix) as a particularly forceful part of his theological argument from design. As he claimed, the eye is surely a superbly fashioned optical instrument. It is also something else, a superb example of maladaptive historical legacy. The retina consists of a series of special layers in the functionally appropriate sequence. A layer of light-sensitive cells (rods and cones) stimulate nerve endings from one or more layers of ganglion cells that carry out initial stages of information processing. From these ganglia, nerve fibers converge to form the main trunk of the optic nerve, which conveys the information to the brain. All layers are served by blood capillaries that provide their metabolic requirements. Unfortunately for Paley's argument, the retina is upside down. The rods and cones are the bottom layer, and light reaches them only after passing through the nerves and blood vessels.

Of course the eye can still play its role as a precise visual instrument. The tissues intervening between the transparent humors of the eye cavity and the optically sensitive layer are microscopically thin. The absorption and scatter of light is ordinarily minor, and functional impairment seldom serious. Yet the fact of maladaptive design, however minimal in effect, spoils Paley's argument that the eye shows intelligent prior planning, and the visual effect is real and routinely demonstrable. Red blood cells are poor transmitters of light, but when moving single file through capillaries can cause only a negligible shading of the light sensors. In larger venules and arterioles they cast dense shadows and blot out images. That we do not ordinarily perceive these shadows is the result of minute involuntary eye movements, which keep the blood-vessel shadows moving, and of our brains recording the flux of images as continuous pictures. The reality of the shadow of the *vascular tree*, and the seriousness of the problem it presents, can be demonstrated with a flashlight and instructions from a visual physiologist.

This is only one of the functional problems related to the inversion of the retina. Another is caused by the optic nerve arising on the wrong side of the sensory layer so that it must go through a hole in the retina to get to the brain. The diameter of the nerve is far greater than that of any retinal blood vessel. That means a large hole, and wherever it is there will be no vision. This is the reason for the blind spot, about 30° right of the point of focus in the right eye, 30° to the left in the left. The visually lateral position of each nerve exit means that the eyes are blind to different parts of a given scene. With both eyes open, we can see everything in the visual field. Our retinal blind spots rarely cause any difficulty, but *rarely* is not the same as *never*. As I momentarily cover one eye to ward off an insect, an important event might be focused on the blind spot of the other.

There would be no blind spot if the vertebrate eye were really intelligently designed. In fact it is stupidly designed, because it embodies many functionally arbitrary or maladaptive features, of which the inversion of the retina is merely one example. These features are there, not for functional but for purely historical reasons. Some affect all vertebrates. Others are more taxonomically limited, merely affecting all mammals, or all snakes, etc. (Goldsmith 1990). The vertebrate eye originated in a tiny transparent ancestor that had no blood corpuscles and formed no retinal images. The retinas arose as light-sensitive regions on the dorsal side of the anterior end of the nervous system. Evolutionary conversion of a flat to a tubular nervous system put the future retinas inside. In subsequent evolution, the photosensitive layer pushed outward from the brain to become part of the complex optical instrument known as the eye. All through history this layer has retained its position beneath the other layers of the retina.

That there is no unappreciated but functionally urgent reason for the vertebrate retina to be inverted is shown by the molluscan retina. A squid has an eye that functions in a way that closely parallels that of a vertebrate. Light passes through a pupil, is focused by a lens, and forms an image on a retina. Unlike that of a vertebrate, the retina of a squid is right side up. Molluscan eyes evolved independently of vertebrate eyes, and show an entirely different suite of historical legacies.

My second example may be less dramatic, but perhaps it illustrates even more clearly an early functional mistake that affected all subsequent history. It is also a geologically more recent and better documented history. Animal gonads, including those of vertebrates, are generally internal organs, although their products, at least those of males, can function only outside the body. So the internal gonads must always have associated tubing that gets gametes or partly developed offspring to the outside. Male mammals, while they conform morphologically to these conditions, are exceptional in one respect. Normal body temperature is too high to permit normal sperm production. It is necessary at some time prior to reproductive age (or prior to a breeding season) for the testes to have their temperatures reduced. For this reason they move from the ancestrally normal (internal) position to a special structure, the scrotum, where they will be separated from the outside by only a thin layer of tissue. Chapter 9 has more to say about the temperature constraint on mammalian spermatogenesis.

This is merely the final episode of a complex story of shifts in organ positions in the abdominal cavity (Booth and Chiasson 1967). The caudad shift of the testis began long before the temperature constraint arose, probably long before mammalian ancestors became warm blooded. This testicular migration brought it closer to the point where the semen is discharged. This should mean, one might think, that ever shorter tubing would be needed as the testes moved closer to the point of discharge. In fact the economy of duct material was frustrated by each testis taking the wrong path to the posterior. It went dorsal to the ureter, which drains the kidney into the bladder, but the sperm duct goes ventral to the ureter. It gets hung up on this other tube, and further movement of the testis imposes a need for a longer, not a shorter sperm duct (Fig. 6.1).

As an analogy I suggest picturing yourself using a garden hose to water some plantings scattered over a lawn. You might move in a rational manner, so as to minimize the number of trees or other obstructions that the hose might catch as you drag it about. Or you can be as short-sighted as natural selection, and head from one planting to another without regard to any ultimate efficiency of time or tubing. Whenever the nozzle fails to reach a destination, you simply add another segment of hose. You could well end up using far more than the needed length, looped

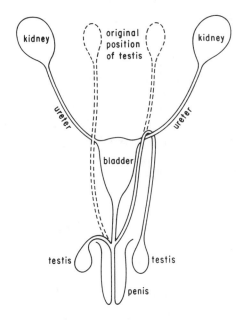

Fig. 6.1. Diagram of the human male urinogential system as it actually is (to the right) as a result of the slow evolution of an ever more posterior position of the testis. To the left is shown what might have been (testis passing ventral to the ureter) if the evolutionary process were capable of anticipating future needs or rectifying past mistakes.

in a complex pattern around obstructions on the lawn. The analogy fails in one respect. The testis had only the one obstruction to negotiate.

The two examples illustrate the concept of phylogenetic legacy. Any number of others could have served. Every organism will show a long list of characters that make no adaptive sense but record past adaptations. The historical explanations proposed can be tested by independent observations on the same or other Recent organisms or from the fossil record. Of special interest are evolutionary steps that can be taken by a wide variety of organisms but, once taken, are not likely to be reversed. Examples are: acquisition of heteromorphic sex chromosomes, loss of the cycle of meiosis and fertilization, exclusive self-fertilization, dioecy, diploidy, haplo-diploidy, polyploidy, a turn of Muller's ratchet (discussed in Chapter 8). This list is from Bull and Charnov (1985) who use the term *irreversible evolution*. Harvey and Partridge (1987) prefer to speak of evolutionary *black holes* and add another item to the list. Once the larvae of a parasitoid evolve lethal weaponry for eliminating competitors, it can never be lost. KILL ALL OTHER LARVAE IN THE HOST is always an ESS that can exclude any associate inclined to peaceful coexistence.

The generation of diversity by cladogenesis furnishes every population with a unique set of historical legacies. In this sense an organism is a living record of its own history. In addition to whatever other values it may have, it has the same value as any other historical document. The loss of the Stellar sea cow and the Adam-and-Eve orchid were the same kind of loss to historical scholarship as the burning of the library at Alexandria. The current wholesale extinction of organisms is especially tragic and ironic because we are only now learning to read history in molecular structure, where the writing may well prove clearer and more detailed than in morphology and other phenotypic end states. To the aesthetic and economic arguments for the conservation of taxa, we should add the argument that the white rhinoceros and the blue whale must be immensely informative books that we have not yet had the skill or the time to read.

6.2 Natural selection and phylogenetic constraint

A neo-Darwinian tenet in conformity with Darwin's original belief is that evolution depends mainly on the accumulation of many minor changes and seldom takes big steps. An implication of this belief, briefly mentioned in Chapter 1, is that adaptation is seriously constrained by phylogeny. Natural selection never designs new machinery to cope with new problems. In an important sense it does not even redesign old machinery. It can only achieve a slow alteration in the parameters of old machinery. Given enough time, natural selection can readily turn something like an arm and hand into something like a leg and hoof, much less readily into something like a bifurcate crustacean appendage. Appreciation of the long-term blindness of natural selection makes it possible to understand why the giraffe, in evolving an ever longer neck, did not add more vertebrae. Adding a new vertebra would have required a redesign of the neck skeleton. A slow increase in one parameter of each vertebra, its length, was the available way selection could bring about the general elongation.

What may seem like a new mechanism for a new purpose, like the human hand with its unique capacity for precise manipulation, is often just a quantitatively altered older mechanism, such as a prosimian paw. A more extreme example of old machinery quantitatively altered for a new function is the mammalian middle ear, an association of greatly modified jaw and hyoid arch bones. This may be pushing the idea of mere parameter optimization rather far. In many respects it makes sense to regard the complete mammalian ear as a new organ with a new design, while recognizing that it arose as an association of parts of old organs: dermal and epidermal structures in the pinna, a hyoid arch element, jawbones, and a specialized part of a piscine ancestor's acoustico-lateralis system. Ultimately any organ has to have a phylogenetically original form, of which later forms are quantitative modifications. One way for a new organ to arise is by association of parts of old organs, and the mammalian ear is an excellent example of this kind of origin.

This idea also has its limitations. Ultimately one must account for the origin of the old organs that supplied pieces for such new ones as the mammalian ear. Mammals had a protist ancestor, perhaps a billion years ago. No quantitative modifications of protist structures or assemblages of pieces thereof can form mammalian organs. The protist-to-mammal idea must deal with the question of the origins of evolutionary novelties and complexities, a much-discussed topic (Buss 1987; Bonner 1988). Here I will venture a generalization: new structures arise in evolution in one of two ultimate ways, as redundancies or as *spandrels* (defined below).

There are many good examples of novelty from redundancy. Serially repeated parts that are closely similar to each other often vary in number in a population without obvious effects on fitness. It would follow that, for an individual with more than the minimum number, one or more of the parts must be functionally redundant or could easily become so. I presume that this was true of the gill arches of some primitive vertebrate. The redundancy allowed two of the arches close to the mouth to become specialized for an originally accidental usefulness in ingesting food. Gill arches gradually became jaws, but the resulting animal still had plenty of gills. In a similar way anterior crustacean legs became jaws, and leaves on stems became reproductive accessories with no photosynthetic ability. In the terminology of Gould and Vrba (1982) such changes illustrate adaptations for one function serving as *exaptations* for a second, and then becoming modified by selection into *adaptations* for the second.

The conversion of a functionally redundant character to a new use must proceed by the blind groping of natural selection for the improvement of current machinery for current uses. Selection does not redesign, it only pushes parameter values towards local optima. Every step of the way, as Darwin (1859, Chapter 6) made clear, had to be immediately useful to each individual possessor. No future usefulness is ever relevant. Yet however weak and shortsighted the process may be, there is no theoretical limit to how far it can go. Ultimately it may produce machinery so different from the original that it may require detailed study for a biologist to recognize the homologies.

The new machinery may also be enormously more complex than the original. A good example is the ilicium of an anglerfish (Grasse 1958, p. 2487). It began as a functionally redundant dorsal fin ray. Now in the anglerfish order it is a complex angling mechanism of great taxonomic diversity. Minimally it is a long filament ending in a lure that can be moved in ways that mimic worms or other prey of the fishes on which an anglerfish preys. Elaborations include complex markings or appendages and even light organs on the lure. At any given moment in evolution, natural selection was merely making quantitative alterations in what had been parts of a fin ray, mostly its numerous separate and partly redundant joints. Ultimately this process produced a mechanism that most observers would recognize as quite different in design and much more complex than the original fin ray.

A spandrel, in the sense of Gould and Lewontin (1979), is a structure arising as an incidental consequence of some evolutionary change. One of their examples is the human chin, which was not evolved for any functional reason, but as a prominence left behind when the dental arcades shrank from protohominid to a modern size. Even if, as seems likely, the structure of the chin is currently optimized as a strengthening element for the lower jaw, its origin as a prominent facial feature was incidental to the reduction of the space required by the teeth. Now that we have chins, and much familial and no doubt genetic variation in them, they can be acted upon by natural selection. Perhaps our future evolution will provide us with much better devices for holding telephones or pillows as we use both hands to work with writing materials or a pillow case.

Or perhaps, given that men tend to have more prominent chins than women, male chins have begun to evolve by sexual selection into structures to rival the antlers of Irish elk. If this seems far fetched, consider the effects of selection on originally modest clusters of feathers on male peacocks, or the evolutionary elaborations of originally simple structures such as the external ears of bats (Walker and Paradiso 1975) or the noses of elephants (Grasse 1955, pp. 752–5) or the Rhinogradentia (Stumpke, 1957).However trivial the circumstances of its origin, the human chin has the potential for endless complexity by parameter optimization, given the requisite selection pressures.

If my view of the origin of mammalian play is correct (pp. 68–9) it would have arisen as a spandrel. Whatever its origin it might, of course, be modified by subsequent selection to serve useful functions. A list of human behavioral and intellectual capabilities must be mostly a list of spandrels, as was pointed out by S. J. Gould (1980, pp. 57–8):

"Our large brains may have originated 'for' some set of necessary skills in gathering food, socializing, or whatever: but these skills do not exhaust the limits of what such a complex machine can do. . . . Built for one thing, it can also do others, and in this flexibility lies both the messiness and the hope of our lives."

I am in entire agreement with this sentiment, but hope that it does not dissuade people from asking what our special braininess was actually selected for. Recent suggestions that selection for algorithms to provide for language acquisition are a key factor in human evolution are surely worth pursuing (Pinker and Bloom 1989). Hopes similar to Gould's, that salvation may lie in our cognitive spandrels, have been expressed by many others (Dawkins 1976; Alexander 1981, 1987; Singer 1981; Williams 1989).

The constraint of gradualism is nicely shown by discrete variables. I presume that some ancestor of the modern bats had typical small-mammal litter sizes, perhaps a mean of five and variance of one. Changes in environment and way of life made it more productive to have fewer young, and evolution took litter size to its ultimate limit of one with near-zero variance (Walker and Paradiso 1975). The early stages of this process were probably easy. A shift from a mean of 5.0 to 4.9 might happen in a few generations with minimal disruption of other characters. A shift from 2.0 to 1.9 is another matter. It requires that single births not be consistently selected against in a population in which twins are the rule. Twins are usual for bears, and even this litter size may form a kind of evolutionary trap from which bear species are unlikely to escape (Ramsay and Dunbrack 1986). The singleton births of bats must be a much more binding trap, perhaps another item for the list of evolutionary black holes (Harvey and Partridge 1987).

The primates are another example. Twinning today is abnormal in all higher primates, and possibly adaptive only in the one species with an obstetrical technology that permits a high rate of survival of twins. Human twinning was maladaptive in advanced societies as late as the eighteenth century (Bulmer 1970; Anderson 1990). It must often be true that low numerical values of characters are evolutionary traps from which escape is difficult. An example suggested by Vitt (1986) is clutch size in lizards. Reproductive investment level for lizards with invariant numbers of young can still evolve by changing egg size, and adult body size can change from any need to accommodate larger or smaller eggs. These are continuous variables for which the low-value constraint would not apply. They can evolve to be in functional harmony with low clutch size, which Vitt regards as resistant to evolutionary change. Numbers of mammalian cervical vertebrae (Chapter 1) would be a similar example. Each individual vertebra is a specialized organ with special relationships to anatomically nearby structures. By contrast, vertebral counts in some fish species can be quite variable (Boetius and Harding 1985). I presume that it makes little functional difference whether an eel has 69 or 70 caudal vertebrae.

Mammalian vertebral numbers and the numbers of young per brood are no doubt characters of critical functional importance. This need not be true of all small-number character states. Darwin (1859, Chapter 14) noted that vestigial characters were generally quite variable in expression. It is probably of little functional significance whether the number of jointed rays in the rudimentary pelvic fins of a sculpin are two, one, or zero (Hubbs and Hubbs 1944).

6.3 Developmental constraints

According to the authoritative treatment of this topic (Maynard Smith *et al*. 1985, p. 266), 'A developmental constraint is a bias on the production of variant phenotypes caused by the structure, character, composition, or dynamics of the developmental system.' Developmental constraints are merely a special kind of phylogenetic constraint, because developmental mechanisms are as much a legacy of past evolution as any other mechanisms. I treat them separately because of the special body of ideas and literature on them (Maynard Smith *et al*. 1985; Arthur 1988; Wimsatt and Schank 1989). My treatment is mainly a commentary on Maynard Smith *et al*. (1985).

I think their definition quoted in the last paragraph perfectly reasonable, but find that they actually include more phenomena than seems justified. In particular, they cloud the issues by including fitness costs as developmental constraints (their 'change, inaccessible for selective reasons,' p. 270). If selection is listed among the constraints, then what is it that is constrained? Developmental constraints would normally be thought of as making it difficult for certain phenotypes to be produced, not as a decrement to their fitness once developed. That there are no normally two-headed vertebrates is not the result of any developmental constraint. The vertebrate developmental system is perfectly capable of producing individuals with two heads, as any nineteenth-century compendium of teratology, or any experienced herdsman will confirm. The persistent one-headedness of vertebrates results from consistent selection against two-headed variants. The unfavorable selection results from the catastrophic functional disruption caused by any mutation or environmental stress great enough to result in two heads on an animal designed for one, and there is no way of gradually evolving the additional head and gradually adapting the rest of the body to two-headedness.

If normalizing selection can always be relied upon to maintain the one-headed character of vertebrates, what about the usual one-tailed condition. A few vertebrates have lost the external tail, and all caudal vertebrae have been lost at least once, in the fish family Molidae (Grasse 1958, pp. 795, 2282–4). I presume that tail loss comes about by gradual reduction of the one tail until it is no longer there. I am not aware that any vertebrate has made the other possible change, from one to two, at least not in nature. Some domestic goldfish varieties have two caudal fins, and the supportive caudal vertebral elements, paired structures in all fish with caudal fins, are laterally separate (Hervey and Hems 1968). This bifurcation of the end of the caudal skeleton, and the two complete caudal fins, might be recognized as the first step in a fantasy of adaptive radiation of two-tailed vertebrates.

Unlike all two-headed vertebrates, slightly two-tailed goldfish are viable, at least in the special conditions of fanciers' aquaria. Might not there be some special conditions in nature where the beginnings of two-tailedness would be selected for in some vertebrate? Perhaps in the near postRecent there will be a seahorse with its prehensile tail ending in two fingers formed by a splitting of the bilateral elements of the final vertebrae. In subsequent evolution the doubling might be extended so that there can be said to be two tails. Are we sure we will have to wait for the postRecent for this? Might not the next issue of *Copeia* announce the discovery of a more-or-less two-tailed seahorse?

Perhaps the normal vertebrate feature of having one heart is more worthy of discussion than one head or one tail. Having two kidneys instead of one has added significantly to many a human life span, and I imagine that an extra heart might do the same. It also happens that a detailed recent study shows that two-hearted birds of various species are produced at rates as high as one in several hundred (Taussig 1988). Yet even if it is granted that a properly optimized two-hearted vertebrate would be superior to its one-hearted ancestor, there is no reason to believe that minor mutational steps in the direction of having two hearts would consistently increase fitness. The adaptive valley between one heart and two is a phylogenetic, not a developmental constraint. The resulting normalizing selection for one-heartedness explains the absence of normally two-hearted vertebrates.

Maynard Smith *et al.* (1985) recognize two general categories of developmental constraint, the *universal* and the *local*. Their recognition of universal constraints seems to be essentially an acceptance of the doctrine of mechanism (Chapter 1). Their main example is conformity to expected relations among lengths, forces, and speeds in systems of levers, such as animal legs. I would prefer to restrict the concept of developmental constraint to those they propose to be local to the taxon, of whatever rank it might be. Their first example, the stem structure of monocots, is a good one. This major division of the seed plants lacks developmental machinery for adding tissues outside those already formed. A result is that tree trunks growing in thickness as they grow in height, as we expect of dicot trees, are unknown among the monocots.

As was noted forcefully by Darwin (1959, Chapter 5), different characters may show correlated variation. If the correlations are among quantitative traits they are now termed *allometry*. Sometimes the correlations arise for obvious physical reasons, for instance the dependence of fecundity on size in fishes (Carlander 1969–77), and sometimes from obvious functional reasons. Character correlations are a kind of character, as subject to variation and selection as any other, and therefore expected to be adaptive (Wagner 1987, 1988). Many examples are obvious, but that does not mean unimportant. In our own species, length of the left leg is positively correlated with length of the right, heavier than average leg bones with heavier than average leg musculature. No doubt this is partly because different components of adaptive machinery are independently optimized by their different genetic bases. It is also true that development can be adaptively cued by usage. The effectiveness of exercise for muscle development is widely recognized. Whatever the developmental details, the body is a complex of coordinated parts, and it is not surprising that one measurement may have considerable value for predicting another.

Other correlations have no obvious explanation, either physical or functional, either because we do not understand the physics, or because they are adaptive in ways we do not understand, or perhaps for some other reason. An example that was not understood at first is the correlation between body size and relative antler size among species of deer (S. J. Gould 1973) (see Harvey and Pagel (1991, pp. 176–7) for a recent explanation). Wherever there is a non-zero correlation between two polygenic characters in a population, no matter what the current developmental or prior evolutionary cause, developmental constraint may be recognized. Successful selection for one character is likely to change the other (the related concept of pleiotropy will be discussed in the next section). In both of the illustrated correlations (Fig. 6.2), the evolution of increased y will be easy if x can decrease without serious cost, but will be difficult (constrained) if x must remain the same or increase, and perhaps very difficult (Fig. 6.2B).

This sort of constraint is no doubt extremely common, especially for artificial selection, which often selects for combinations that have not been adaptive in the past. With tightly constrained allometry there may be scarcely any progress with the character selected for, but another may be unexpectedly altered to a striking degree. An example was found by Francis (1984), who selected one lineage of male paradise fish to win conflicts with other males, and another lineage to lose. After four generations there was little difference between selected and control lines in contest outcome, but the winner line produced a great excess of male offspring, the loser line a great excess of female. A practical example of character correlation opposing artificial selection is found in the poultry industry (Lerner 1953). Egg productivity is negatively correlated with viability in highly selected hens. Stock improvement is frustrated by the

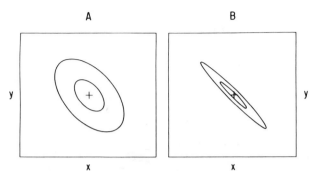

Fig. 6.2. Hypothetical patterns of variation of two quantitative characters, *x* and *y*, in a population. The + shows the mean for both *x* and *y*, and the lines show contours along equal densities of points, each point representing the pair of measurements for a single individual. Inner and outer lines might be one and two standard deviations from the means. The same arguments would apply to positive relations (allometry) between *x* and *y* as to the negative ones shown here. A illustrates a moderate, and B an extreme example of developmental constraint.

potentially most productive hens being impossible to keep alive. In nature, rates of long-term evolutionary change are such that the directional selection responsible must be far weaker than that often imposed on farm or laboratory stocks. The constraints illustrated (at least in Fig. 6.2A) need not seriously impede the progress of evolution by natural selection (Zeng 1988).

Maynard Smith *et al.* (1985) recognize that the absolute facility with which selection can produce effects may be less important than relative facility. If a moderate intensity of selection can increase both *x* and *y* by 1% per generation, and both *v* and *w* by only 1% in ten generations, major changes in both pairs of characters could appear instantaneous in the fossil record. The rate difference would be important only if the two sets of changes oppose each other in some way. It must often happen that changed conditions challenge an organism with a new kind of problem to solve (new predator or pathogen to avoid, new resource to exploit, etc.). There may be more than one kind of change that could provide the required capability. Whichever kind is most easily begun and immediately adaptive would be the one most strongly initiated. Once the modifications needed for one kind of solution to the new problem are well started, alternative suites of modification would lose any advantage they might have had, and might be selected against if they interfered or competed in any way with the adaptation actually being evolved. In this way slight differences in character correlations might trigger one of several potentially adaptive modifications and constrain the entire future evolution of the affected lineage. Noirot and Pasteels (1988) use this kind of argument to explain differences between termite and hymenopteran social systems.

I argued above that character correlations, like other characters, must often be adaptive. Other correlations represent physically inevitable trade-offs; large size gives a mating advantage to a male platyfish, but is achieved at the cost of delayed maturity (Van Vorhees 1988). Inevitably also, some correlations must arise incidentally from the developmental machinery and are likely to be adaptively neutral in historically normal environments. If the environment changes, such correlations could act to retard or prevent some kind of adaptive modification. Maynard Smith *et al.* (1985) are perhaps inconsistent on the relative frequencies of adaptive and nonadaptive correlations. They maintain (p. 267) that 'covariance among traits . . . is subject to selection,' but then later (p. 269) that adaptive randomness is generally expected of developmental constraints.

It is often expected that alterations of early developmental processes are likely to have manifold continuing effects. The earlier the alteration, the more likely would be major later disruptions. It follows that changes late in ontogeny should be easier to evolve than earlier ones, and that early developmental stages would tend to remain conservative, even in groups with greatly diversified adult forms. Many examples of conformity to this expectation can be cited, and they form the basis for *von Baer's law* (Futuyma 1986, p. 303). Formal models of the developmental basis for von Baer's Law have recently been explored. Arthur (1988), Kauffman (1985), Wimsatt (1986), Wimsatt and Schank (1989) have modeled adaptive evolution as changes in a developmental process of branching causes and effects. In any such models, early alterations have more effects on the final state than later ones, and it is not surprising that changes in early life-history stages in such models prove difficult to achieve and that von Baer's law is thereby explained.

I suggest that this modeling shows something else: development can not be modeled realistically as a system of branching causation, with what Rice (1990) calls *burden* and Wimsatt and Schank (1989) the *generative entrenchment* of early stages. The developmental program must be more like modular computer code with many self-contained parts that are forbidden to affect other parts, except for specifically programmed exchanges of information. Examples of drastic metamorphosis, with earlier structures abandoned and later ones formed anew, may just be easily appreciated examples of a process that permeates all development. The protective insulation of different parts of the developmental machinery can not be perfect, but it must be far more effective and prevalent than can arise from nonmodular systems of cause and effect. This view of development is in fact widely accepted (Parks *et al.* 1988; Sachs 1988;

Wagner 1989; Wray and McClay 1989. The concept of developmental *networks* (Maynard Smith *et al.* 1985, p. 268) and Bonner's (1988, p. 174) *gene nets* embody the same idea. Bonner attributes the modular-program idea to Jon Seger.

I also suggest that von Baer's law, and the reasoning and evidence on which it is based, have been taken too seriously. The exceptions are simply too many and too diverse and too striking. Consider the degree of morphological uniformity of adult bivalve mollusks and the great structural diversity of their young: specialized pelagic larval stages in many marine forms, the strikingly different structures of brooded young, the parasitic specializations of larval unionids. Coe (1949) recognized this effect in various other mollusks, with seemingly similar and closely related adult forms having morphologically divergent young. The same can be said of beetles (Peck 1986) and sea urchins (Sinervo and McEdward 1988). In various fishes the larval and early juvenile stages can be strikingly polymorphic, while adults are much more uniform (Hubbs and Whitlock 1929; Paine and Balon 1986; Marliave 1989).

Various closely related echinoderms may differ in whether the larval stage is a plankton predator, with all the elaborate machinery needed for this way of life, or develops entirely on yolk reserves (Parks *et al.* 1988; Sinervo and McEdward 1988). Strikingly different larval forms may characterize subspecies, or be found in the same population (Levin *et al.* 1991). Echinoderm larvae may be more diverse in molecular characters than the adults (Wray and McClay 1989). It seems that early life history stages are just as free to evolve special modifications for special conditions as are adults, and can do so without serious effects on adult morphology. This conclusion echos Thomson's (1988).

6.4 Genetic constraints

A genetic constraint on adaptive evolution is any quantitative or qualitative deficiency of genetic variation (Loeschcke 1987). Complete absence of additive genetic variability in a character would be the simplest sort of genetic constraint on selection for that character and is commonly encountered in artificial selection. The mean of an experimentally selected character may change appreciably for a few generations but then reach a value at which further selection produces no further change, as in Johannsen's (1909) classic study of bean size. Additive genetic variability may be exhausted, and whatever genes have been favorably selected are now fixed or at frequency-dependent equilibria. Any remaining phenotypic variability is environmental or interactive in origin. Intense and prolonged selection for desired attributes of domestic animals and plants, such as egg productivity of poultry, may be similarly frustrated by exhaustion of additive genetic variation (and probably by opposing natural selection, as noted above). It is also true, as noted by Darwin (1859, Chapter 1), that domesticated animals and plants can still respond strongly to selection for economically important characters. A five-fold increase in the fat content of maize kernels is a recent example (Crow 1988).

This kind of constraint can seldom be important in nature, except for populations so small and isolated that genetic drift can fix some allele at nearly every locus. Any population that persists long enough to experience appreciable and prolonged shifts in character optima ought to be able to track the changing demands. If its original store of genetic variability is not adequate, it can often be expected to acquire such variability by mutation or gene flow. Calculations and data summaries by Lande (1975) and Lynch (1988) indicate that mutation alone can augment variability much more than would be necessary for quite rapid evolution as seen in the fossil record. Wild populations seldom show prolonged directional change at even one per cent of the rates seen in initial phases of response-to-selection experiments. Prolonged directional selection on quantitative characters in nature (other than major fitness components or *winnings*, as discussed in Chapter 5) must normally be far weaker than that imposed by breeders.

While it may often be true that an evolving lineage has abundant variability for further evolution, it is sometimes not true. Selection for altered sex ratio in diploid organisms with genetic sex determination is sometimes effective (e.g. Francis 1984), but usually not (Charnov 1982, pp. 114–20; Clutton-Brock and Iason 1986). There often seems to be no effective variability in the cytological machinery of meiosis and fertilization. Secondary adjustments of sex ratio, such as sex-biased abortion or infanticide, could be opposed by selection rather than constrained by genetics. They might be too costly for parents, or opposed by selection on embryos for resistance to any such parental adaptation.

Maynard Smith and Sondhi (1960) note that artificial selection for any sort of left–right or male–female difference is generally ineffective. Attempts to select *Drosophila* for long left wings and short right wings, or for long-winged males and short-winged females, will make little progress. All the familiar one-locus variants normally studied by geneticists, for instance the eye-color mutants of *Drosophila*, affect both sides and both sexes the same way. They would provide no basis for selection for asymmetries.

But there are many examples in nature of evolved left–right or male–female asymmetries, and some good examples of the coordinated evolution of both. The major cheliped of a male fiddler crab is different from the other on the same animal, and different from either on a female. A more extreme example is the tusk (hypertrophied upper right incisor) of a male narhwal (Best 1981; Kingsley and Ramsay 1988). I presume that variation in nature is just as unfavorable to selection for asymmetry as it is in laboratory stocks, but that nature is patient and can make do with a severely restricted supply of variation. A one per cent change in a hundred generations would necessarily be considered absence of evolution in an experimental population. It would be a meteoric rate of evolution in the fossil record. Degrees of responsiveness of different characters to artificial selection are unlikely to measure the readiness with which natural populations alter such characters over geological time, as noted by Ricklefs and Marks (1984).

The tendency for mutations to have similar effects on left and right sides and in both males and females is a special case of the more general phenomenon of pleiotropy, which can be defined in various ways. I use it here to mean a one-locus developmental constraint. The pleiotropy of two quantitative characters can be modeled in the same way as any developmental constraint that links the magnitudes of two characters (Fig. 6.2, p. 83). Pleiotropy of a new mutation is less apt to be adaptive than a multi-locus allometry, which must have been selected for in the past or at least permitted by selection. Pleiotropy is also less likely to constrain response to selection, because the different mutations affecting one character need not have similar effects on another. Nature can make adaptive modifications in one character by using only those genes that require the least onerous compromises for other characters (Lenski 1988).

6.5 Unity of type and Bauplan

Related to the ideas discussed above, especially to phylogenetic constraint, are the concepts of *unity of type* and *Bauplan* (Gould and Lewontin 1979; Valentine 1986). Members of any major group of organisms: primates, gymnosperms, deuterostomes, etc., are said to share a basic body plan, and it is supposedly difficult for the evolutionary process to alter this plan. This idea is misguided and dispensable.

It is certainly true that for any monophyletic group it is possible to find suites of diagnostic characters by which it can be identified. Without them no such group would be postulated in the first place. It is equally true that the higher the taxonomic rank, the fewer will be the features shared by all. The most inclusive groups can be recognized only by use of profoundly subtle homologies. Comparative morphology had to reach an advanced state before it could be seriously proposed that acorn worms and acorn woodpeckers are both members of a phylum Chordata, which would exclude both earthworms and butterflies. Having one head, one tail, and one heart seem to be normal conditions for vertebrates but, as argued above, there is no reason to think of them as parts of a developmentally inflexible *Bauplan*. Another example was pointed out by Ruud (1965). The text-books would have you believe that all vertebrates have hemoglobin, but there is a diverse group of vertebrates, the Antarctic fish family Chaenichthyidae, in which hemoglobin is entirely lacking. This is merely one of many possible examples of generalizations for which exceptions must be recognized. All tetrapods have two lungs, except for snakes, which have but one, and plethodontid salamanders, which have none (Storer and Usinger 1965).

If there are certain characters that have steadfastly resisted change for many millions of years, and others that change readily, then the unchanging ones would collectively constitute a real *Bauplan*. The null model for this idea would be a phylogeny with random changes in one or a few of a long list of characters in every daughter lineage (such random phylogenies are discussed in Chapter 9, and by Felsenstein (1988)). If all but 10% have changed after a 100 million years, it is expected that all but 1% would have changed after 200 million. If real phylogenies resemble this random model, there is no *Bauplan*. If the list of unchanged characters is greater than would be expected of exponential decay, perhaps with moderate variation in probability of change for different characters, the *Bauplan* idea would be supported. Until there is some such evidence I would accept the *Bauplan* idea only in the trivial sense that every bifurcation in a cladogram separates two clades with two different *Baupläne*. Any taxonomically significant branching must result from the acquisition of at least one apomorphic character in at least one of the branches, and therefore represents a departure from an ancestral *Bauplan*.

7

Diversity within and among populations

Individuals in a single population can differ from each other from a variety of causes, some discussed in Chapter 5. The present chapter will use these and a few additional ideas to discuss variation within populations, and then go on to the broader question of differences among populations and phylads at all levels of divergence.

7.1 Natural history as the foundation of comparative biology

Suppose we got the following results of an experimental test of strains S and s on diets D and d. S produces phenotype Φ on both diets, but s produces Φ on diet D and ϕ on diet d (Fig. 7.1). Simple field observations could yield the same kind of data: One species has the same phenotype in habitats H and h, while another has one phenotype in H and a different one in h. A frequent reaction to this kind of information is to

	D	d
S	Φ	Φ
s	Φ	ϕ

Fig. 7.1. Hypothetical test of strains S and s on diets D and d.

say that one form (S) is developmentally inflexible while the other (s) shows phenotypic *plasticity*, with the implication of adaptive adjustment to the special demands of environmental factors, such as diet d as opposed to D.

All this of course is nonsense, unless there is some independent understanding of how φ is adaptive in relation to d. The tabulated results could refer to one strain producing a normal phenotype on two different diets and the other a normal one on one diet and scurvy on the other. No conclusion about adaptation can be established merely from the kind of data shown in Fig. 7.1. They would not distinguish between an effect of the environment and an organismic response that prevents the environment from having an effect. An adaptive interpretation of data like those shown in Fig. 7.1 must always be bolstered by some understanding of how the phenotypic modifications would act to increase fitness.

In classic cause–effect analysis it is sufficient to say that, for strain s, D caused Φ and d caused φ. If the analysis goes no further it misses a most fundamental biological distinction. Effect φ could be a nonadaptive change from d's interference with development, or it could be a response of the organism designed to prevent d from having some detrimental effect. A response always requires the prior existence of an evolved mechanism for sensing the environmental factor or some reliably associated stimulus. The sensor must activate machinery that then acts to prevent some effect of the environmental factor. A major direct effect is never adaptive or, at least, is as unlikely to be adaptive as a major mutational change. If environmental input causes adaptive change, it can be assumed that the change results from an evolved response mechanism. The cause–effect analysis must always go on to the question of whether the observed effect is caused by environmental interference or by the organism acting in response to some cue.

Fortunately there are abundant examples of independent understanding of the adaptive significance of phenotypic differences between environments. For instance we can imagine how different kinds of dentition and jaw structure would adapt a fish to different diets. Different prey offered to young of the fish *Cichlasoma managuense* cause them to develop feeding structures adapted to the diet experienced (Meyer 1987). This species could be represented by strain s in Fig. 7.1. Fish fed consistently on microcrustacea and those fed on coarser insects can both grow rapidly. Another species, specialized for an insect diet and incapable of altering its mouth parts for plankton, might grow rapidly on an insect diet but be stunted on even abundant microcrustacea. Its stunting would make it another sort of example of strain s in Fig. 7.1.

An adaptive response to a diet (as a cue) need not be an adaptation to the diet. A kind of food item may merely be predictive of some other environmental factor, and a dietary cue may trigger adaptation to this other factor. The lower row of Fig. 7.1 could record the fact that *Nemoria arizonaria* caterpillars look like catkins when raised on a diet low in secondary compounds and look like twigs when these dietary components are abundant (Greene 1989). Their responsiveness to chemical cues is so adjusted that they look like catkins in the spring when catkins are abundant and like twigs after the catkins fall. An appreciation of cryptic coloration gives us an understanding, independent of the data in Fig. 7.1, of how it would augment fitness to have this seasonal change in visual appearance. The catkin form and twig form are more divergent than many distinctions used to separate species, but are achieved in the absence of any genetic difference. There are many other examples of distinct and clearly adaptive phenotypic differences achieved by environmental cues during development, in the absence of genetic differences. The castes of social insects provide many striking examples (Jeanne 1988; West-Eberhard 1989). Others are discussed below.

Clark (1991) advocates use of *plasticity* only for results of developmental switch mechanisms, such as response to diet at a critical stage of development. She suggests *flexibility* for readily reversible characters, like the color of a chameleon. I suggest that *plasticity* be abandoned entirely, at least for adaptive responses. It implies molding by an environmental constraint, like the conformity of an oyster shell to the contours of a rocky crevice, rather than an active biological process. If an environmental cue evokes what is thought to be an *adaptive response* it should be so identified, and the nature of the postulated adaptation explained. If not, such neutral terms as *variation* are adequately descriptive.

The important point is that the mere fact of environmentally correlated variability should never be assumed to imply adaptation. It might indicate the opposite, environmental interference with development. For observed differences to be recognized as adaptive they must conform to expectations based on a prior understanding of the way of life of the organism. There are innumerable examples of such conformity to adaptive expectations, for instance in the phenomena of inducible defenses (Crowl and Covich 1990; Harvell 1990; Wilbur and Fauth 1990).

7.2 Causes of variation among individuals

Much of ecology and evolutionary biology depends on the sort of procedures outlined above for interpreting differences among individuals in a population. The following is meant to be a complete list of causes contributing to differences between two individuals.

(1) *Ontogenetic stage*. Stage differences may be conspicuous, as between

a tadpole and frog, or slight, as between early and late second-instar larvae. This is a classic kind of taxonomic challenge (e.g. which tadpole goes with which frog?). It can be a source of frustration to population ecologists when only adult and later juvenile stages are readily identifiable, as is often true of tropical insects, fishes, and flowering plants.

(2) *Age.* They may be of different age, a common reason for being different in stage, but they could be of the same age but different stages. If so, the one in the more advanced stage probably has higher phenotypic fitness. As a general rule, the faster the development the higher the fitness, because the longer it takes to mature, the more opportunity there is to die first. Exceptions occur when development is adaptively synchronized with environmental cycles. If dawn is the optimum moment for hatching in a *Drosophila* population (Pittendrigh 1958), too rapid development would be a liability.

(3) *Frequency-dependent strategy.* They may be employing different life-history strategies maintained by frequency-dependent selection. Sex differences are the most commonly appreciated example, but many more are known, many only recently discovered. Within a sex there may be more than one reproductive strategy, like territorial *hooknose* salmon and the dwarf parasitic *jacks* (Gross 1985). Single populations may have two or more trophic niches, as in the bluegill discussed below and various other fishes (Sage and Selander 1975; Riget *et al.* 1986; Meyer 1987; Noaks *et al.* 1989). Batesian mimicry polymorphisms are another striking example (Turner 1988; Davies and Brooke 1989). The clearest examples of frequency-dependent selection are these discrete polymorphisms, but high levels of continuous variability can also be maintained by such selection, as discussed in Chapter 5. For this and the next source of variation below, the different forms may be genetically fixed for each individual (like Batesian mimics) or triggered by environmental cues (like the choice of sex from the size and sex of an associate in the polychaete worm *Ophyrotrocha* (Sella 1984; Monahan 1986).

(4) *Optimized response.* They may show individual adjustments to environmental factors other than the strategies employed by associates. Such responses could result in such polymorphisms as winner vs loser strategies that may depend on how well an individual is doing in its current efforts to achieve reproductive success. Qualitative phenotypic shifts in whole populations may be caused by response to environmental change. The arrival of a predator, for example, may cause a whole population of microinvertebrates to grow spines or other defensive weaponry (Washburn *et al.* 1988; Harvell 1990). More commonly the differences between individuals would be merely quantitative and reflect different optima for different circumstances. Adaptive variation in skin

pigmentation in a closely related human group may be largely a function of the extent of recent exposure to sunlight.

(5) *Genetic load.* One or both individuals may suffer from one or more kinds of inherited defect. Mutation and migration can introduce locally maladaptive genes, and sexual reproduction generates maladaptive genotypes from adaptive genes. The commonest kind of genetic load may be homozygote disadvantage, but multilocus effects are possible, especially for progeny of individuals adapted to different habitats. Extreme examples of influxes of locally maladaptive genes can be found in hybrid zones, which Barton and Hewitt (1985) term *hybrid sinks* for incoming genes (see also Slatkin (1987)).

(6) *Epigenetic load.* They may differ in effects of environmentally inflicted trauma or deficiency (relative to some ideal that may be seldom experienced): poorly healed or unhealed injury, parasite load, inadequate nutrition, trace-nutrient scarcity, culturally inherited low social status, deception or manipulation by a competitor or exploiter. Fitness differences between monozygotic twins would arise entirely from epigenetic load, even if they chose different developmental pathways.

(7) *Neutral variation.* Slight differences from mutation, recombination, gene flow or environmental effect would be expected to have slight effects on fitness, and if slight enough can be considered negligible. Much variation in molecular characters is often considered neutral, as discussed in Chapters 2 and 4. Variation caused by differences in strategy or individual adjustment can be neutral in a special sense. At a mixed-strategy ESS the alternatives are all of exactly equal fitness. Two individuals with different responses to different stimuli can also be equally fit.

Because of varying genetic and epigenetic load, individuals may differ in winnings (current fitness), as discussed in Chapter 5. The most common indication of a fitness difference would be a size difference, the larger individual of an equal-age pair having higher fitness. Solbrig and Solbrig (1984) review the close relationship of size to fitness in plant populations. A similarly strong dependence of fitness on size must be true for many invertebrates and fishes (see discussion of bluegill sunfish below). Less fit individuals may actively assume behaviors or developmental programs different from highly fit members of the population. An extreme example would be sterile caste membership in a social insect. A worker diet for a larval female bee inflicts a great epigenetic load. As a result, her fitness will be higher through the worker than the queen mode of development. She can strive for the propagation of her genes as they are represented in her mother's and sisters' germ cells and her mother's stored sperm.

Perhaps she may produce a few sons by laying unfertilized eggs. At best, she will achieve, as a worker, far less reproductive success than a queen might, but far more than if she had tried to be a queen after having been reared on a worker's diet.

7.3 Examples of variation within a population

Some variation within populations is within individuals, in two senses. Modules or ramets of a clone or genet can differ from each other for any of the nongenetic reasons listed above. For instance some members of a *Daphnia* clone may form sexual eggs while others do not, some may be sick while others are healthy, etc. Even within a single unitary organism there may be noteworthy differences between paired parts. Such differences may often result simply from imperfect canalization, but they can also be adaptive. One of a lobster's chelipeds is heavier and rather like a pair of pliers, the other more like scissors. Whichever cheliped first gets more than a threshold amount of use early in life becomes the heavier (crusher) appendage (Govind 1989).

It may be that only human populations are sufficiently well known to provide many clear examples of all seven of the sources of variation listed in the last section, but in a few respects our species may be quite atypical. All mammals are exceptional in showing no strong relation between size and female fecundity, and size would also be but weakly indicative of human fitness in other respects. If the various professions and life styles are considered facultative life-history strategies, as seems reasonable, we would provide an extreme example of largely facultative, frequency-dependent polymorphisms, although the male–female difference may be the only morphologically striking genetic polymorphism maintained by frequency-dependent selection. The immense literature of clinical medicine can be considered a documentation of genetic and epigenetic loads. The same source documents variation in current winnings, in any assessments of health and fertility.

Another good example of a richly variable and well-studied organism is the bluegill sunfish, *Lepomis macrochirus*, a common centrarchid of broad distribution in temperate North America. More detailed information than is presented here can be had from the works cited below and those reviewed by Carlander (1969–77). Eggs are guarded in nests by territorial males. They hatch, at least the luckier specimens, into yolk-sac larvae about 5 mm long. After a few days they are independent and actively feeding on zooplankton. By the time they are about 25 mm in length, when the sexes are not yet externally distinguishable, the growing juveniles have differentiated into two trophic morphs specialized for different habitats (Layzer and Clady 1987; Ehlinger and Wilson 1988; Ehlinger

1989, 1990). A deeper-bodied form keeps to the weedy shallows inhabited by earlier stages where it must feed increasingly on insect larvae. A more elongate form, with smaller mouth and finer scales, moves to more open waters where it continues feeding mainly on plankton.

In the following year another strategic divergence shows itself (Gross and Charnov 1980; Dominey 1981; Coleman *et al.* 1985; Ehlinger 1990). The females and most of the males continue to grow, but with no sign of maturation. The other males begin to mature, and may be ready to spawn at two years. When they do they will be at a small fraction of the size of territory holders, and they will parasitize these much older males. As a female spawns with a territorial, the smaller males will dart from a nearby hiding place into the nest and shed semen, perhaps successfully fertilizing some of the eggs, despite vigorous attack by the nest owner. The cuckolder's young will then be tended and defended along with the young of the territorial.

After a season of this parasitism by stealth, the early-maturing males will continue to grow through the third summer. The following spring they will spawn parasitically again, but with different tactics. At age three they are likely to be larger than the ideal size for stealth and may attempt to spawn by mimicking the form and color and behavior of females. Territorial males may be deceived enough to court the female mimics and allow them a close approach to the nest, and thereby the opportunity to intrude quickly when a territorial male spawns with a real female. The parasitic form continues thereafter as a female mimic, but is not likely to live more than a very few spawning seasons.

Males that do not mature at age two continue to grow until they are big enough to compete for territories, at about eight years of age. These holdings are usually quite small and closely associated with other males' territories to form a localized colony. Interior territories are less affected by egg predators and are held by the largest males (Gross and MacMillan 1981). A successful territorial male may fertilize most of the eggs of several females in a breeding season and be far more successful than any cuckolding parasite, but the delay in maturation has a high price. A male using the territorial strategy is much less likely to survive to maturity than is a parasite.

The difference between parasitic and territorial males is genetically determined, at least partly (Mart L. Gross, personal communication), as is the analogous jack-hooknose contrast in sockeye salmon (Iwamoto *et al.* 1984; Gross 1985). It is not yet known how the sexual strategies may relate to the juvenile feeding strategies. At evolutionary equilibrium, territorial and parasitic males must have equal average fitness, as must males and females. In both cases, at least one form needs the other, and excess abundance of either form would decrease that form's fitness. If territorials were overabundant they would devote more of their energies to fighting each other and courting females, and this would allow a higher average success for the parasites. If parasites were in excess, they might be expected to arouse the territorials to greater anti-parasite vigilance and aggression, and the parasites would be competing with each other for vulnerable territorials. They would thereby have their mean fitness reduced below that of the territorials. These contrasting male sexual strategies provide a good illustration of a mixed ESS maintained by frequency-dependent selection. The differences in size and proportions between the two kinds of male bluegills are fully comparable to those used to diagnose different species of fish. Comparable male dimorphisms are known in many species (reviewed by Van Vorhees 1988). Fig. 7.2 shows one example, and others are discussed in Chapter 8.

There is only one strategy for female bluegills. They visit the nesting sites and spawn preferentially with the larger, dominant males in central territories. They mature at about four years and commonly survive through more spawning seasons than most males. Like the males, they

Fig. 7.2. Representative specimens of two male morphs from a single population of the Poeciliid fish *Brachyraphis rhabdophora*. Growth in this group of fishes is determinate, and neither morph grows after sexual maturity. The genetic basis and action of selection on these morphs are under investigation (Reznick *et al.* 1992). Photograph by David N. Reznick.

are highly variable in growth rate and in size at a given age. Fecundity is strongly size-dependent, with eggs making up about 10% of body mass. Carlander (1969–77) lists egg counts varying from 3820 to 29,769.

At any given age, individuals of both sexes are quite variable in size and other measures of fitness. Small size may result from any of the factors 3–6 in the last section. Whatever its cause, reduced size may mean delay in reaching maturity, and therefore greater risk of dying without issue. The account above is greatly simplified and deals mainly with averages determined in a few careful studies in a few of the thousands of bodies of water inhabited by bluegills.

There is no reason to believe that the variability of the bluegill is unusual. Biologists have long been aware of the prevalence of variation among individuals of a single population (Darwin 1859, Chapter 2). The new awareness is of the role of natural selection in maintaining such variation, either through frequency-selected polymorphisms or facultative optimization. Admirable reviews of the phenomenon of adaptive variation within a population and of its evolutionary significance are available: Clark (1991), Clark and Ehlinger (1987), West-Eberhard (1987, 1989), and D. S. Wilson (1989). Some recent examples published too late for inclusion in the reviews are a trophically significant dimorphism of bill structure in a finch (T. B. Smith 1990), a switch between filter-feeding and endoparasitic life history by a protozoan in response to presence or absence of a predator (Washburn *et al.* 1988), and Walls and Ketola's (1989) assessment of the cost of an inducible defense in a cladoceran.

Even when it is obvious that a variable character must affect fitness, it is not always easy to understand the role of natural selection in its evolution. A good example is clutch-size variation in the snow goose (*Chen caerulescens*) in the colony at La Perouse Bay, Manitoba. During his many years of study at this locality, Fred Cooke and his associates (Cooke *et al.* 1984; Rockwell *et al.* 1987; Cooke 1988) have amassed an extraordinarily rich mass of data on the ecology, behavior, and life history of this bird. The relationship between clutch size and breeding success is highly variable from year to year, but on the average over 20 years, larger clutches have a higher net production of young. This greater productivity is apparently not entirely balanced by long-term costs to either parents or young. It is also observed that clutch size has significant heritability (about 0.20). These observations lead us to expect a trend towards increased clutch size over the period of the study, because response to selection supposedly equals the product of selection coefficient and heritability, or about 0.066 eggs per generation (Cooke *et al.* 1990). In fact, average clutch size has decreased in 20 years by about 0.20 eggs (Cooch *et al.* 1989).

Expectations on such matters as response to selection almost always depend on various simplifying assumptions, and a common reaction to a failure of an expectation is to question one or more of these assumptions (Lakatos 1978; Mitchell and Valone 1990). Simple models of response to selection assume constant conditions, but environmental change may often result in phenotypic change, such as the decreased clutch size reported by Cooch *et al.* (1989). This possibility is explored by Cooke *et al.* (1990). The breeding colony has been increasing in size, and this could result in food shortages, which could depress clutch size. A more subtle possibility is that the population, very much a part of the environment for every member, may be evolving an increased competitive ability. This would be expected if the apparent selection on clutch size is really selection for ability to compete for resources. With total resources as a limiting factor for the population, there would be no way for an increasing competitive ability coded in the gene pool to express itself as larger mean clutch size. This sort of genotypic change over time, reversed by an opposite and more-than-equal environmental change, may be an example in the temporal dimension of *countergradient selection* (de Jong 1987) over a habitat gradient.

The power of resolution of field data will often be too weak to provide more than a crude estimate of the trade-offs and net selection on any character related to fitness in a natural population, although selection can often be detected and sometimes roughly measured (Endler 1986). It must also be expected that discrepancies between observed reproductive success and expectations from ESS or optimization models will be common, for reasons discussed mainly in Chapter 5. There can never be any assurance that the observations were made in a really normal habitat for a representative period of time. Conformity to *a priori* design specifications remains the preferred method for demonstrating the long-term operation of natural selection. It is surprising that human socio-economic behavior in various twentieth century societies is as biologically adaptive as is commonly observed (Betzig *et al.* 1987; Rasa *et al.* 1989). In the terminology of Chapter 5, the compass of human behavioral adaptations can not be infinite, but it is surprisingly broad.

7.4 Cladogenesis

Widespread appreciation of the prevalence and magnitudes of social and ecological specialization within single populations is a rather new development. There is nothing new in the idea that there can be appreciable adaptive radiation among populations of what are considered single species, especially if they range over large areas of diverse environments. Darwin (1859, Chapter 2) attributed the geographic variation prevalent in widespread species largely to selection favoring somewhat different character states in different climates and communities. The same sorts of phenomena, described in the terminology of Mendelian genetics, provided important evidence for pioneers of the neo-Darwinian synthesis (Dobzhansky 1941, Chapters 5 and 6; Mayr 1942; Huxley 1942; Stebbins 1950, pp. 42–52, 118–20). More recent studies of geographic variation are likely to emphasize behavioral (Houde 1988; Dowdey and Brodie 1989; Foster 1992), physiological (John-Adler *et al.* 1988; Parsons 1982; Brodie and Brodie 1991), life history (Billington *et al.* 1988; Reznick 1989) or molecular (Prosser 1986; Nevo 1986) specializations.

Mayr (1966) and Endler (1977, 1986) give comprehensive reviews of the phenomena and detailed studies of the patterning of variation among localities and of its historical and ecological interpretation. Their reasoning is largely based on the premise that divergence leading to taxonomic distinction requires at least the partial isolation of gene pools within a species. The isolation may arise from barriers to dispersal (allopatric speciation) or from mere distance (parapatric speciation). Both Mayr and Endler argue that sympatric speciation from disruptive selection within a locality is uncommon.

Whenever one gene pool becomes two, by processes discussed by Mayr and Endler, or for any other reason, a cladogenetic event (*cladogeny*) has occurred. Most cladogenies are of trivial importance in the history of life. If I extract one or more pairs of *Drosophila* from a culture and use them to establish another culture, one gene pool has become two, but the event is not likely to be worth anyone's close attention. Cladogenies in nature can also isolate populations for trivial periods of time on an evolutionary time scale. A flood may turn sections of a river into oxbow lakes and thereby isolate small populations of riverine organisms in ecologically novel habitats. A similar flood a few years later may reunite such populations with ancestral stocks in the river.

Serious study of the production of descendant from ancestral gene pools is more justified if the gene-pool separation is likely to last a long time. It is still more worthy of attention if the descendants have already become genetically different in important ways and include large numbers of individuals spread over large ranges. Many such cladogenies may occur within what would be generally recognized as single species, and the constituent populations can be arranged on dendrograms like those routinely used for separate species or higher taxa. A within-species cladogram may need to depict both branching and anastomosis of lineages, as in Fig. 2.1B (p. 14).

The importance of the study of phylogeny within a single species follows as an extension of Cracraft's (1989) arguments for a phylogenetic species concept as the appropriate basis for the study of phylogeny in general. In my view he ought not to have restricted his argument to populations that have diverged sufficiently for their members to be confidently identified. I have no quarrel with his equating this origin of reliable diagnosis with the origin of species. I would object only to the implication that subspecific and lower levels of distinction are inappropriate for phylogenetic analysis. In this I concur with Templeton, Wilson, and other contributors to Otte and Endler (1989).

Interpretation of phylogenetic diversity, in relation to adaptive evolutionary change, is a complex topic that has received detailed attention from some distinguished contributors, such as G Bell (1989), Gittleman and Kot (1991), Harvey and Pagel (1991), and Ridley (1983). Thoughtful discussions of molecular diversity are provided by Koehn and Hilbish (1987) and Nevo (1986). Here I will deal only with a few special aspects of phylogenetic diversity, mostly related to my conviction that any comparison between groups, whether it relates to a major difference or no difference at all, needs to be reconciled sooner or later with natural selection. There is no need to suppose that such differences must be directly favored by selection. They could represent variable costs of the variable development of some other differences that are adaptive. They could arise from genetic drift (Lande 1976), in which case it would be appropriate to inquire why the changes were permitted by selection. Whatever their genesis, they can be studied initially as phylogenetic patterns without adaptive interpretations.

7.5 Comparative biology without functionalism

As critics of the adaptationist program have pointed out, natural selection is not the only evolutionary factor, even at a strictly microevolutionary level. It is equally worth emphasizing that there is no lineage in which natural selection ever fails to operate, at least in the material as opposed to codical domain (Chapters 2 and 3). This encourages skepticism of claims for neutrality or adaptive irrelevance in any character, whether morphological or molecular. The assumption that underlies molecular clock determinism, that much of molecular divergence is caused entirely by mutation pressure and genetic drift, has always been vulnerable to some general criticism, e.g. from Wallace (1989), and contrary data. Much of molecular diversity can be shown to conform to patterns expected of natural selection, and the responsible environmental factors can sometimes be identified (Koehn and Hilbish 1987; Watt *et al.* 1983). 'It does appear that whenever allozymes are examined closely, significant functional differences are found among alleles' (Endler 1986, p. 158).

None of this denies that some variation must be neutral or that a wish to understand adaptation need be the only reason for studying diversity. Given that taxonomic differences arise by descent with modification, a maximally accurate and detailed description of that history is a quarry well worth pursuing. I take it that this, among other goals, is what phylogenetic systematists are striving to reach. It is possible that their success could be enhanced if they are unburdened by prior convictions about the causes of the evolutionary changes they try to discover. Unless they succeed, between-group comparisons made to enhance understanding of adaptation will be frustrated by errors or uncertainties on actual genealogies. It also helps to know the time scale of phylogenetic events. The idea that horse dentition was evolved for its usefulness in grazing on grasses can be ruled out if it arose before the grasses did.

7.6 Functional interpretations of phylogenetic divergence

There are three traditional methods of comparative biology for demonstrating the adaptive significance of various features of organisms. The first is to consult an intuitive appreciation of the features themselves. We think we understand why feet are adaptive for terrestrial locomotion and fins for aquatic. I refer readers to the Appendix for classic examples of intuitive appreciation of the features of organisms. The second is to note the effects of the difference on adaptive performance, a matter dealt with in Chapter 5. The third method is to note the facts of association among organisms' features or between their features and their habitats and ways of life. We note that feet are commoner than fins for terrestrial vertebrates and fins commoner among fishes. Here I will deal critically with this third method of comparative biology.

Suppose we have reason to suspect that some phenotypic character y functions in such a way that it should show lower values as some environmental factor x increases. Investigation then shows an x–y association as in Fig. 7.3A. The regression of y on x is significant for our five points, and we thereby claim empirical support for our hypothesis of decreased y being adaptive in relation to increased x. Is this a valid conclusion? I suspect that the answer may depend, for many investigators, on just what is represented by the data points. Suppose the prediction is that territory size in some bird population will be larger in open-field habitats than in woodlands. The plotted x-values might be canopy height, and y may be measured in hectares. In this case I would expect general agreement that the data confirm the prediction and thereby support the adaptive explanation.

But suppose the data represent, not measurements on individual organisms and their habitat variables, but species means. Suppose further that the upper-left and lower-right clusters are regarded as separate

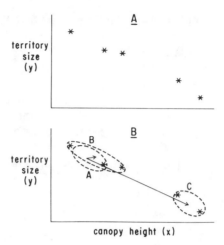

canopy height (x)

Fig. 7.3. A shows hypothetical values of environmental variable *x* and organism feature *y* supposedly adaptive to *x*. B shows the same data as mean values for species *a*, *b*, and *c* in genus B, and species *d* and *e* in genus C. Cladistic analysis (B) yields the connecting branches (arrows) deriving from an ancestral genus A with *x* and *y* values close to those of genus B. This lower diagram (B) shows a top view of a dendrogram of the genera and species derived from ancestor A. Current practice would rule out this phylogenetic pattern as support for an adaptive relation between *x* and *y*.

genera (B with three species, and C with two), and that cladistic analysis indicates the relationships shown in Fig. 7.3B. Ancestral genus A is inferred to have been rather similar, in habitat and territory size, to Recent genus B. Now I expect an authoritative challenge to the idea that the data support any sort of prediction (G. Bell 1989; Pagel and Harvey 1988; Harvey and Pagel 1991). The cladogram shows only a single significant example, the origin of genus C, of an expected evolutionary change. Genera A and B provide one observation, genus C a second, and no such two-point relationship can possibly be significant.

The original assumption (Fig. 7.3A) was that there were five independent data points observed in a study of five developmentally independent birds in a single population, but perhaps this is wrong. Further study might show that the population has a phenotypically distinctive single-locus dimorphism at locus *z*. The *zz* individuals have mouths and foraging behavior suitable for seizing insects on the wing in dense woodlands, and they always nest in such habitats and need only small territories. Genotypes *Zz* and *ZZ* have mouths and foraging behavior suitable for picking insects from low vegetation in open fields, where they always nest and demand large territories. The observations (Fig. 7.3A) can be explained entirely by a single difference, the z-locus dimorphism. One such favorable observation can hardly be cited as evidence for any proposition. Or can it? I am unaware of any study of the logical differences among phylogenetic independence of gene pools, phylogenetic independence of gene lineages within a single gene pool, and developmental independence of different individuals in a population or clone.

But is evolutionary change in the predicted direction the only valid comparative evidence for adaptation? I suspect that the likely answer depends on what the question really means. If we are asking about a proposed cause of an evolutionary change, the number of changes is the number of relevant observations, and Fig. 7.3B would not be convincing evidence that a change in x causes a change in y. If we are asking about the maintenance of associations in independent gene pools, the data support the dependence, whether presented as in Fig. 7.3A or B. At least this would be true if there is reason to believe that y is currently responsive to selection. In principle this could be resolved by experimental investigation.

If my hypothetical organisms were small mammals rather than birds, and the dependent variable coat color instead of territory size, experimentation might be more feasible. For any example, responsiveness to selection would show that the stability of observed associations must result from normalizing selection, and that the number of pertinent observations is the number of independent gene pools in which this stability is maintained. For these reasons the number-of-changes requirement may be too stringent for the recognition of adaptation from phylad comparisons. My position assumes the general prevalence in nature of stability from normalizing selection and of responsiveness to directional selection.

For example, it seems unrealistic to assume that the size similarity of current representatives of *Homo* and *Pan* is to be explained entirely by their having some Pliocene ancestor of about the same size. It is well known that *Homo* has an appreciable heritability of size and is subject to normalizing selection for size (Van Valen and Mallin 1967). There is every reason to believe the same of *Pan*, and of other sexually reproducing organisms that are not closely inbred and have not recently experienced strong directional selection for the character under consideration (exceptions to this free-to-evolve rule are discussed in Chapter 9). Harvey and Pagel (1991) make it clear that similarities among closely related phylads can result from similarities in selection pressures and need not imply any kind of phylogenetic constraint, but the implications of this concession for their main argument are not developed. Also they quote with approval a statement of Felsenstein (1985b) urging that attention to phylogeny is indispensable for comparative biology.

Early in their work, Harvey and Pagel (1991) assert, correctly, that appreciation of the functional significance of any attribute of any organism can only arise from consideration of alternative attributes. Their illustrative attribute is the whiteness of the fur of a rabbit in a snowy environment. To appreciate the whiteness we must 'inevitably make an implicit or explicit comparison with rabbits that are not white' (Harvey and Pagel 1991, p. 13). All subsequent discussion relates to one kind of *explicit* comparison, between phylogenetically divergent forms.

I would assert that if only one rabbit had ever been observed, and it was uniformly white and seen on a snowy field, the observation would justify viewing the whiteness as adaptive camouflage. An imaginary brown rabbit on the same field would be the relevant comparison. The camouflage inferrence at this stage would be quite tentative, but an effective basis for further research on conformity to design specifications. There would be no way of knowing whether the whiteness is obligate or facultative or how elaborate an adaptation the camouflage really is, but we could imagine elaborations and test for them. If the habitat is snowy for most of the year, whiteness would be adaptive most of the time, but we would imagine that a change to some other color for snowless seasons would make the coloration more adaptive. This and many other refinements of the idea of adaptive crypsis could be tested. The proposed refinements would arise from an imagination informed by Darwinism and a knowledge of natural history and relevant constraints. Knowing that a rabbit is clothed in a metabolically dead and slowly renewed fur would make it unlikely that a field biologist would look for rapid chameleon-like matching of background color. Various tests of adaptive performance could also be carried out, for instance by dyeing some rabbits a conspicuous color and noting the effect on predation rate.

My position is that the demonstration of conformity to design specifications is superior to phylogenetic comparison as a way of demonstrating adaptation. The phylogenetic comparisons also have great value in other respects. They can demonstrate evolutionary cause and effect when we have reasons to anticipate it and, more importantly, when we do not. If some y often and unexpectedly decreases as some x increases in some group of organisms it may identify some unappreciated adaptation. It would stimulate a search for some undiscovered adaptive machinery of which x and y are components.

I also believe that the comparative method, as developed by Harvey and Pagel (1991) and their forerunners, may be profitably extended. We need a body of theory that can deal with equal effectiveness with comparisons among phylads in a dendrogram, among individuals in a Mendelian population, among ramets in a clone, and within physiologically defined individuals. I see no reason to worry about independence of evolutionary changes in considering whether it is adaptive for people to be more darkly pigmented in the summer than in the winter or, for those who drive American or Continental cars, darker on the left arm than on the right.

8

Some recent issues

In this chapter I discuss five miscellaneous issues on which I believe my opinions to be a minority view among contemporary biologists: the lek paradox, the female pheromone fallacy, *Schreckstoff*, the helpful stress effect, and abuses of the species concept. My model for this and the final two chapters is Darwin's (1859) Chapter 6 ('Difficulties of the Theory') and Chapter 7 ('Miscellaneous Objections to the Theory of Natural Selection'). With hindsight we can see that Darwin's chapters successfully surmounted most of the difficulties and answered most of the objections. I expect no comparable success, but hope that I can at least convince some others that evolutionary biology today, no less than in Darwin's time, faces some serious anomalies and conceptual challenges.

8.1 The lek paradox

Discussions of sexual selection have proliferated in recent years, along with a growth of awareness of its importance. A useful review and initiation into the spirit of current debates is found in Bradbury and Andersson (1987). One of the controversies is on the evolutionary basis of mate choice by females in species with males that make no contribution to the rearing of offspring. A clear example is seen in birds in which males form leks, localized groups where they compete with each other for positions of dominance. Females arrive singly at leks and are courted by the males, with highly variable success. Often one of the males of a dozen or more will win the acceptance of most of the females, and one or two other males will get almost all the rest (Bradbury *et al.* 1985). For most males the mating effort will be entirely fruitless. Females once mated return to their own individual nesting sites. There are good recent reviews of the lek phenomenon by Bradbury and Gibson (1983), Gibson and Bradbury (1986), and McDonald (1989).

The most challenging aspect of lek mating systems is female behavior, especially the strength and near unanimity of preferences for mates. Females may spend much time and effort in fending off the attentions of subordinate males while the dominant is busy with another female. When he finishes, the next female in line readily accepts him as a mate. Once inseminated she leaves, goes to a remote nesting site, and raises her brood of eggs and chicks with no assistance from their father.

Prior to the development of modern population genetics, an intuitive good-genes theory seemed a plausible explanation for a female's choosiness: she tries to mate with the fittest male, thereby providing her offspring with genes for high fitness. We now know that this would not be effective. Any genes that reliably confer high fitness must be already fixed in the gene pool. Fitness varies mainly from such factors as interactive genetic and environmental effects that have little heritability (Falconer 1981; Gustafsson 1986; Charlesworth 1987; Gibbs 1988). If males will provide no assistance in raising offspring, and if one male's genes are as likely as another's to confer high fitness, there seems to be no reason for females to be so discriminating. Hence the *lek paradox* (of Borgia (1979)).

Efforts to solve the problem have relied mainly on linkage disequilibria between genes for competitive traits and those for female preference (Fisher's runaway process), or on the finding of previously unappreciated sources of additive genetic variation in fitness. I have no doubt that both approaches have met with some success. Both indicate advantages to females in choosing high-status males. Unfortunately they may not, even jointly, provide an adequate explanation for the degree of development of female choice and of its strength in driving the evolution of competitive male weaponry and adornments. For example, various sources of fitness heritability (mutation pressure, gene flow, etc.) were analyzed by Taylor and Williams (1982). They showed that there would be only trivial differences in offspring fitness between a female with perfect ability to choose the fittest mate and one that mated at random. For comparison, consider what would happen if random choice were substituted for normal selection of food or habitat.

The unimportance of any good-genes effect is currently controversial. It is supported by general arguments against any appreciable heritability of fitness (Charlesworth (1987) and other references above) and against the *sexy son* hypothesis, a special form of the good-genes idea (Kirkpatrick 1985). Efforts to demonstrate fitness enhancement by female choice have given negative results for flour beetles (Boake 1986), cockroaches (Moore and Moore 1988) and indigo buntings (Payne and Westneat 1988). Other studies show positive effects, for *Drosophila* (Partridge 1980; Taylor *et al.* 1987), for a seaweed fly (Crocker and Day 1987) and for a cricket (Simmons 1987).

My conviction that good-genes models and other current explanations for female choice are inadequate rests mainly on intuitive misgivings, but the continuing efforts of many able workers to refine and evaluate the models suggest that I am not alone in my uneasiness. Gibson (1988), in reviewing Bradbury and Andersson (1987) suggests that the modeling fails to explain why females seem to be choosiest in species in which all males are alike in ability to rear offspring (all of exactly zero ability). Perhaps another approach is needed, and I sense some promise in a recent suggestion by West-Eberhard (1984), especially in relation to two other ideas, asymmetrical fitness distributions (Chapter 6), and the venerable ethological concept of *supernormal stimuli*. West-Eberhard suggests that females with no mate-choice inclinations can hardly be expected to exist. Any animal is programmed to react adaptively to stimuli, and different kinds and intensities of stimuli can be expected to produce different reactions. Whatever the stimulus–response patterns of females might be, a male can be expected to try to manipulate them in his own interest.

Suppose there were a bird population with males and females of generally similar appearance and (*reductio ad absurdum*) no special courtship signals. Males ready to mate simply approach any adult females in their home ranges and try to mate. Unfortunately for the females, males have a cheap and gradual development of mating behavior, and even some that are not fully fertile will attempt to mount females. Also perhaps, males of a related species with a somewhat different eye color may make mating attempts. We would expect under these conditions that females would evolve some discrimination: accept only full-size (apparently adult) males with the right eye color. If so, we would expect the males to attempt to appear large enough and to have the right eye color. Both characters would be subject to exaggeration in fully adult males, but would pass through all intermediate stages in juvenile male development.

Obviously a runaway process is now initiated, with males becoming (or trying to look) ever larger and with ever more obviously correct eyes, and females escalating their demands for male size and eye display, to assure that a mate is properly fertile. There need be no special genetic mechanisms at work, such as linkage disequilibria between genes for female choice and those for male size and ocular display, as forms the core of sexy-son and related models. There is no special advantage to a female in picking a male of unusually large size or with brighter than average eyes, as is needed for the Hamilton–Zuk model of parasite-driven sexual selection (Hamilton *et al.* 1990). It is merely necessary to assume that female choice has an asymmetrical fitness distribution (Chapter 5). Picking a male much smaller than average or with markedly off-color eyes can have a serious cost. I assume that selection for the avoidance of clearly inadequate males will inevitably reward males that provide supernormal stimuli.

If one male happens to have some vaguely eye-like spot somewhere on his plumage, this may incidentally enhance the eye-pattern perceived by females and slightly augment his mating success. Further escalation of female demands may make the eye spot a common and ultimately a necessary feature. If what looks like an extra eye is an aid in the mating competition, why not two? and then three? and so on. Ultimately we arrive at *Polyplectron malacense*, with males markedly larger than females and with as many as 40 ocelli in the plumage display, all closely resembling the actual male eyes (Davison 1983).

Size and eyes are two arbitrary characters that were exaggerated by sexual selection in one species of pheasant. Any other character that would serve to identify conspecific adult males might be affected by this process, for instance vocalization early in the history of passerine birds. Adult male fertility and courtship singing develop cheaply and gradually. Females are vulnerable to unproductive matings, with possible wastage of expensive eggs, if they are exposed to subadult males of inadequate fertility. Fagen (1981, p. 23) notes the prevalence of 'presong' in birds, varied fragments of the adult song often repeated by juveniles or by adult males out of the breeding season. Any female in my hypothetical population who is deceived by a presong may lay one or more unfertilized eggs and waste all or part of her reproductive investment.

Under these circumstances, females will be selected to choose males that have songs (or plumage ocelli) of *at least* some threshold degree of development (Fig. 8.1; see also Fig. 5.3, p. 67). The choice mechanism

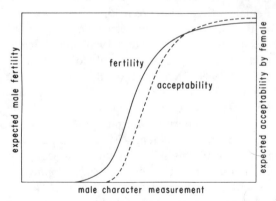

Fig. 8.1. Proposed distributions, at evolutionary equilibrium, of a character predictive of male fertility (solid line) and of female preference for it (dashed line).

will necessarily be imperfect and variable, sometimes causing a deficient and at other times an excess choosiness. Choice of a deficient male may bring a severe fitness penalty, but choice of a male with a more than adequate song will not. The female choice mechanism expected to evolve would more often reward a male with above-average than one with below-average song development. This directional selection would soon result, on an evolutionary time scale, in the fertile, fully adult males having louder or more frequent or more melodic songs. Females in this population would shortly be preferring far more elaborate songs than I originally postulated to be adequate. The same argument can be applied to any male feature (size, ornamental markings, social status) correlated with fertility.

The same argument can also be applied whenever any reaction to a stimulus has the sort of asymmetrical distribution of pay-offs shown in the figure. A preference for supernormal stimuli will soon result. A female will prefer the most conspicuously masculine of her suitors for the same reason that she will prefer to incubate the most conspicuously egg-like object near her nest, even if this means abandoning her own egg for one of exaggerated form and size (J. L. Gould 1982). I am not aware that anyone else has used the ideas of supernormal stimuli and asymmetrical fitness distributions to resolve the lek paradox, but some recent discussions certainly hint at it (West-Eberhard 1984; Burley 1986; Houde 1988; Kirkpatrick 1987; Lande 1987).

It is not surprising that maximum female responsiveness might be elicited by male signals greater than any actually found in the population (supernormal stimulus), as has been shown in the widow bird. Males with tail plumes grossly exaggerated by Andersson's (1982) experimental manipulations were more successful in courtship than were any merely normal males. Females of the viviparous killifish *Xiphophorus pygmaeus* prefer males of the more extremely ornamented (but allopatric) *X. nigrensis* to those of their own species (Ryan and Wagner 1987), and female sticklebacks prefer exaggerated models to any males of their own population (Rowland 1989). My postulated choice mechanism would always lead a female to choose a male with statistically extreme ornamentation or weaponry or social status. The fitness benefit would not be the acquisition of superior paternal genes for offspring, but an increased assurance of providing them with enough paternal gametes.

A testable implication of this view of female choice is that partial or complete sterility, among males with moderate development of secondary sexual characters and of willingness to mate, is of common occurrence. There should be a negative correlation between the development of these characters and the incidence of sterility. Recency of copulation in lekking birds should be a less reliable warning, of reduced fertility, than underdevelopment of display characters. Absence of these relationships in any population would be evidence against my suggestion.

8.2 The female pheromone fallacy

In most animals the rate of production of young is set by female capabilities, and male reproduction is possible only with successful competition for the use of these capabilities. This general basis for sexual selection explains why males are usually more active in courtship and more competitive among themselves. Courtship may in fact be a complex program of male–female interactions, but the males take the lead. They initiate the process, and females may or may not respond in ways that stimulate further interactions. Exceptions are expected and found in those rare groups, such as polyandrous birds (Graul *et al.* 1977) in which essential male services limit the rate at which females reproduce.

There also seems to be another group of exceptions, characterized not by adaptive adjustments to the economics of reproduction, but merely by the medium in which the courtship signals are given. Males initiate courtship by visual, sonic, tactile, electric (Hagedorn 1988) and other non-olfactory signals, but if the initial signal is olfactory it is normally the female that initiates courtship. The world is full of males displaying to females with bright colors and loud songs and conspicuous actions, and of females displaying to males with odors. This is strange.

Or, more likely, wrong. I suggest that female sex pheromones generally do not exist or, at least, are no more common than female initiation of courtship by singing or by waving ornamental fins or feathers. I can hardly dispute the phenomena on which the sex-pheromone concept is based. Male moths do indeed respond to chemicals from reproductively ready females. They detect these chemicals in amazingly low concentrations, fly upwind when they do, and thereby locate the ultimate resources on which their reproduction depends (Farcas and Shory 1974; Schneider 1974). My dispute is with the claim that these chemicals are signals.

Pheromones are 'substances used in communication between members of the same species' (E. O. Wilson 1975, p. 231, under 'Chemical communication'). If the communication concept is to have any special meaning, beyond the general one of response to environmental stimuli, it must imply special signaling machinery that would not be there except for its usefulness in communication. As argued by Burghardt (1970), Otte (1974), and Lewis (1984), a stimulus should be called a signal only if it is normally adaptive for the sender and produced by machinery attributable to selection for the proposed communicative function. The song machinery of a male bird is a good example, and there can be no doubt that his song is a form of communication. The same can be said of the alarm calls of both sexes and the red epaulets of a redwinged blackbird and the red belly of a male stickleback. Likewise for the alarm pheromones of termites and the fragrance of lilies. All are normally of benefit to the sender of the signals, and all are produced by machinery clearly understandable as signaling mechanisms.

But what of female pheromones? The female moth gets a reproductively essential benefit from the male response, but a benefit need not imply an adaptation. The chemical stimulus that recruits the male is a communicative adaptation only if it is produced by machinery designed by selection to produce that response. There is little evidence that this is generally true for supposed female sex pheromones. The adaptation in most species could be entirely male. It is very much to his advantage to be able to discriminate a conspecific female ready to mate from one that is not. He can be expected to use any and all available stimuli that provide this information and enable him to locate the relevant female. To the male moth, a female ready to reproduce smells different from one not ready, and from anything else in his environment. This effect could be produced by a female pheromone, but this can only be decided by examining the female.

Such an examination usually reveals nothing that can be interpreted as a signaling device. Female moths in general do not have taxonomically specific attractants but rather a mixture of substances with various relative concentrations (Roelofs and Carde 1974; Ritter 1979). These substances are produced in minute traces, in contrast to the abundant productions of genuine pheromones, such as the occasional male sex stimulants (Jacobson 1972; Birch 1974) and the alarm substances of termites (Evans *et al*. 1979). Studies of the supposed female-to-male communication emphasize the extreme development of male abilities to detect female odors, not of any female machinery for releasing them (Carde and Charleton 1984). The same seems to be true for other female animals traditionally viewed as releasing pheromonal attractants. Male lizards (Cooper *et al*. 1986), snakes (Mason *et al*. 1989), and salamanders (Roudebush and Taylor 1987) may respond to odors from the genital opening of a reproductive female, but there is no special machinery for producing the odor. The olfactory stimuli are in glandular secretions inside the cloaca and only passively find their way out. It is understandable that various phases of gametogenesis and other aspects of reproductive physiology would affect cloacal secretions in ways that males could detect.

There are many analogies in the realm of visual stimuli. A male stickleback is attracted to a female, in part, because of her dull gray color, which contrasts markedly with the bright reds and blues of his sexual competitors. No one would say that the female is gray so that she can attract a male. The color is explicable entirely on the basis of the foraging and anti-predator values of crypsis. The same can be said for male abilities to tell gravid females from the immature or newly spent. The rotund belly of a female is attractive to a male and elicits a response that may benefit the female, but her rotundity results from distention by eggs, not from any special signaling device. If it were found that some females with few eggs will imbibe water until they are as swollen as those with many eggs, communication (of misinformation) would be indicated. Male fathead minnows feign robustness by imbibing water as they deplete their fat and protein reserves (Unger 1983).

There are in fact a few good examples of female sex pheromones, and they clearly illustrate the essential features missing in the invalid examples. The female scent marking behavior of a tamarin must certainly be regarded as communicative behavior (French and Cleveland 1984), but some closely related species do not show scent marking. There is a female moth that gives off a pulsating spray of attractant from a special organ (Krasnoff and Roelofs 1988). A female spider may produce a substance attractive to males in response to their prolonged absence (Watson 1986). She has a special behavioral pattern by which she applies the substance to her web. If this attracts a male, who mates with her, he may then consume the web so as to avoid the attraction of additional males. This certainly looks like communication by the female and an attempt by the male to frustrate it when it would attract a competitor.

As expected, there are many valid examples of male sex pheromones, which can function in various phases of courtship but usually not in long-distance attraction (Jacobson 1972; Birch 1974). They are emitted by special mechanisms at special moments in the courtship process. Various male insects, for example, have sternal glands that can be everted to give off a pheromone and then retracted to end its release. Others have antennal glands used at close range on females. These are all special structures with abundant secretions and not incidental consequences of the readiness of the reproductive system itself. The compounds produced are taxonomically specific, and not mixtures of substances in somewhat different proportions in different species, as they usually are in females (Linn *et al.* 1987; West-Eberhard 1984). There can be no doubt that these are evolved mechanisms of communication by males to females.

8.3 Schreckstoff

Pasteur's maxim that chance favors the prepared mind is nicely exemplified by one of Nobel laureate von Frich's (1938) chance discoveries. He found that the odor of an 'alarm substance' (*Schreckstoff*) released from club cells in the epidermis of a minnow causes other individuals to show a fright reaction. Subsequent research showed the generality of the phenomenon for three orders of teleost fishes. Upon perceiving the alarm substance, a fish immediately shows its usual reaction to danger (closer schooling, flight, hiding, etc.). For many years these observations were naively interpreted as showing a form of communication whereby an individual attacked by a predator warns its fellows of danger: 'the fright reaction . . . is an important insurance against predation. The alarm substance does not protect the individual, but serves for the protection of its school' (Pfeiffer 1962).

The production and release of alarm substance and the response to it can be considered separately with respect to descriptive details, ontogeny, phylogenetic distribution, and adaptive significance (Pfeiffer 1962, 1977, 1982; R. J. F. Smith 1982). Club cells are larger than most other epidermal cells, usually well below the surface (ca. 100 μm deep), and without any opening to the outside. They have uniform staining properties indicative of chemical uniformity wherever found. The substance is produced and stored in these cells and escapes as a result of mechanical injury to the skin. Club cells and alarm substance appear early in juvenile life, e.g. at 31 days in the zebra danio studied by Waldman (1982), and from their first occurrence can stimulate the fright reaction, which appears later in development (e.g. 50 days in Waldman's danios).

Both club cells and fright reaction are found in most members of the orders Gonorhynchiformes, Cypriniformes, and Siluriformes, which comprise the phylad Ostariophysi. The possession of club cells and reaction to their contents can certainly be considered primitive characters in this phylad, and their absence apomorphic. A small minority, among about 180 ostariophysine species studied, lack the reaction, and a smaller minority also lack club cells. Club cells temporarily atrophy in some North American minnows in the breeding season (Smith and Smith 1983). Males in these species often abrade themselves when digging oviposition sites or from the weaponry (nuptial tubercles) of rival males. If they kept normal club cells the release of alarm substance could be expected to frustrate courtship.

Pfeiffer's (1977) thorough review establishes that loss of the fright reaction (but retention of club cells) is shown by: (1) cave dwellers, (2) piranhas, (3) electric catfishes and gymnotids, and (4) catfish species with heavy defensive armor. These diverse exceptions suggest that the substance must have a function in addition to (or other than) the warning of conspecifics. Otherwise it should have degenerated in all species without the fright reaction. We would also expect the list to provide hints on the ecological factors that lead to loss of the reaction.

Adults of cave fishes are often the largest animals in their communities and in little danger from predators. This would not be true for the young, which might be attacked by adults of their own or other species or by large aquatic arthropods. Some piranha species may reach several kilograms and may or may not be the largest predacious fish in their habitats. They would certainly be vulnerable to such predators as crocodiles and anacondas, and the young must have many piscine predators. Reviews and most of the original reports (e.g. Pfeiffer 1962, 1977; Markl 1972) often fail to specify sizes of experimental subjects and imply that adults were used. The possibility that the fright reaction is missing only in adults of large species would be worth investigating. Loss of the fright reaction by fishes with strong electric defenses or heavy armor also suggests that lack of predator vulnerability causes the loss. This is dubious for the smaller gymnotids, which have electric discharges adequate for sensory and social functions but not for warding off predators.

Retention of the fright reaction by the great majority of the Ostariophysi is clearly adaptive. Injury to a conspecific must often imply danger to oneself, and the smell of the alarm substance from a fellow's injured skin may sometimes be the only way of knowing about the injury. The behavior should be especially effective against an ambush predator in waters where plant growths impede vision. A persistent odor of alarm substance near the predator in a weedy pond might warn many potential victims of the ambush. The acceptability of this as an adaptive explanation is compromised by the lack of association of fright reaction with habitat. It characterizes a single phylad with a great diversity of habitats and life styles.

The adaptive significance of the reaction to alarm substance may be clear, but that of its production and storage is not. Can we really accept the conventional view that it is a mechanism of communication? Is the alarm substance an underwater analog of the sonic alarm calls of birds and mammals? Numerous studies of these avian and mammalian calls have mainly supported kin selection as the factor responsible for their origin and maintenance (Barash 1982). Kin selection is less likely to be implicated for the Ostariophysi. Most species abandon their fertilized eggs immediately, and the minority that tend the young in nests do so only briefly. The young then disperse from the nest and apparently from each other. Most fish schools are temporary groupings that scatter at night, and membership can change from moment to moment (Shaw 1978).

The traditional view that alarm substance is produced for a communicative function is dubious for several reasons that collectively justify a search for an alternative: (1) the persistence of club cells in species in which they do not serve to warn other individuals of danger; (2) the inappropriateness of an olfactory signal in flowing waters, where it can act only in a limited downstream sector; (3) the ineffectiveness of any olfactory signal, in calm aquatic environments, for sending information for which rapidity of transmission should be important (E. O. Wilson (1975, p. 235) notes that diffusion in water is only about a thousandth as rapid as in air); (4) the release of the supposed signal only when the skin is broken, rather than when a predator is perceived, as is usual with sonic or visual alarm signals; (5) the frequent evolution of rapidly transmitted visual and sonic signals by the Ostariophysi and by fishes in general (Protasov 1966); and (6) the difficulty of accounting for such a mechanism by natural selection.

Two alternatives have been suggested, that the alarm substance may be distasteful and repel predators (Williams 1964, pp. 377–8; Verheijen and Reuter 1969), and that it may protect injured skin against pathogens (R. J. F. Smith 1982). It is understandable that a substance strongly distasteful to a predator might be detected in minute traces by olfaction and come to be used as a danger cue. Unfortunately only meager anecdotal evidence supports the anti-predator interpretation (Pfeiffer 1962) except possibly for cannibalism.

Bernstein and Smith (1983) present experimental evidence against the anti-predator theory. They found that trout will feed as eagerly on male minnows with fully developed club cells as on those with the cells atrophied by hormone treatment. They believe that the anti-pathogen idea has greater promise, and note a resemblance in molecular structure between the alarm substance (hypoxanthine-3(N)-oxide, according to Pfeiffer (1982)) and various antibiotics. If the anti-pathogen theory is correct, it may be that even a crude minnow-skin extract would inhibit bacterial or fungal growths *in vitro* and provide the beginnings of a resolution of this difficulty. I expect the problem to be resolved soon, and that we will then have an understanding of *Schreckstoff* fully supportive of current evolutionary theory. If not, it presents evolutionary biologists with a serious anomaly.

8.4 The helpful-stress effect

Long ago Demerec (1951) showed that some streptomycin-resistant strains of bacteria were not only competitively inferior to ancestral strains in the absence of the antibiotic, but were unable to grow at all in its absence (for a recent review, see Saunders 1984). This was perhaps the first clear recognition of what might be called a *helpful stress*. I presume that streptomycin is synthesized by the fungus as a way of combating bacterial pathogens and competitors, but in Demerec's experiments the lethal toxin seemed to become an essential nutrient for bacteria. Similar effects have been or could be recognized in a wide range of biological phenomena.

Mammals may fight respiratory infections by coughing and sneezing, but such host defenses may come to be needed by some target organisms for spreading to new hosts. They might even be manipulated by the pathogen to make them more effective in such dispersal. Many higher plants make special toxins as a way of discouraging herbivores, but then some herbivorous insects may need their favorite plant's chemical weaponry for their own defense or as a way of identifying the plant.

Mammalian herbivory provides a recent and controversial example of a helpful stress for plants. For many years field workers have been making observations that suggest a greater productivity and fitness for grazed plants than for ungrazed (McNaughton 1983, 1986; Belsky 1986) but many remained unconvinced. Recently Paige and Whitham (1987) produced decisive experimental evidence that the effect can be real. Removal of more than 90% of the shoot of the herbaceous annual *Ipomopsis aggregata* as it starts to flower resulted in greatly increased reproductive success from what Paige and Whitham call *overcompensation*. Removal of the stem is followed by the growth of several stems that collectively produce many more seeds than the original would have produced if left undisturbed.

Few biologists nowadays believe, with Wynne-Edwards (1962), that an organism subordinates its own reproductive efforts to the interests of its population. They believe that it tries to achieve the greatest reproductive success it can, given the materials and information available to it. It therefore seems mysterious that a plant capable of making several stems like one removed by a grazer would not do so without the removal. Perhaps the plant's need for grazing is no more mysterious than a bacterial need for an antibiotic, though it may seem so to those whose intuitions are more compelling for mechanical than for chemical stresses.

I suspect that the ungrazed plants in Paige and Whitham's experiment suffered from inadequate information. They did not perceive that their environments would allow the productive use of several stems. I propose that the *Ipomopsis* life-history strategy is based on the expectation that development will often take place under a particular sequence of stresses: crowding often followed by loss of the original shoot. I presume that it commonly experiences this sequence of events in its normal habitat and season of growth. If its stem gets bitten off, so have other stems in its immediate vicinity. Loss of the shoot is the signal that another kind of shoot growth would be adaptive. When it grows in the absence of these normal stresses it may achieve lower levels of reproductive success than we can imagine it achieving with a different growth habit.

I would predict that the responses to grazing found in *Ipomopsis*, and consequent increased fitness under grazing, will be found in other species of similar structure in similar habitats. Other species, for instance those with stronger and larger single shoots, will not be benefited by having these shoots cut off. The limited data available for other species suggest that the extreme overcompensation found in *Ipomopsis* may be rather uncommon (Doust 1980; Belsky 1987; Hendrix and Trapp 1989; Maschinski and Whatham 1989).

In a sense the development of most organisms takes place under an expected level of stress, at least in the form of resource limitation. Birds that develop from larger eggs or in less crowded nests are better nourished and achieve independence with a larger size and greater fitness (Klomp 1970). It is also commonly realized that an organism can be too well fed. The human developmental machinery is not designed to make maximally adaptive use of nutrient levels prevalent in today's wealthier societies. Perhaps, as a general rule, an individual of any species is designed to make effective use of resources one or two standard deviations above historically average abundance. Still higher resource levels will not result in greater fitness, and eventually a decreased fitness is to be expected, as exemplified by human health hazards associated with obesity.

It is normally adaptive for any animal to minimize mechanical stresses and energy wastage, i.e. to avoid injury and be as lazy as circumstances allow. In our present abnormal environments a normal level of laziness, and especially the extreme abnormal levels imposed by prolonged immobilization after serious injury or surgery, may result in degenerative structural changes in skeletal and muscular systems. Normal childhood development of these systems, and of the dental apparatus, may require stresses provided in abundance by the sorts of usage demanded by any normal environment. A world dominated by classrooms, television screens, and mashed potatoes could well be an inadequate source of such stresses. I have more to say on the developmental hazards of our current abnormal environments elsewhere (Williams and Nesse 1991).

The discussion above is hardly a solution to any kind of problem. More serious thought and no doubt some formal modeling is needed on what I call *helpful stress*. More detailed experimental and comparative work on the beneficial-grazing phenomenon might be especially enlightening. If there really are organisms that have increased reproductive success when deprived of resources, in the normal range of resource availability and cues from their environments, they would provide a serious challenge to current understanding.

8.5 Species fallacies

The best known attempt to standardize the species rank throughout biological classification is Mayr's (1942, 1988) *biological species* concept, which defines a species to include all populations that are reproductively compatible, or that would be if they were in contact. Within a species there are no effective *intrinsic* barriers to gene flow. Thus the small-mouth black basses of Lake Opinicon never seem to cross with the closely similar large-mouth black basses in this lake. The two kinds of black bass thereby show themselves to be different species. It is equally apparent that the small-mouth black basses in this lake do not interbreed with those in Lake Wobegon, but this can be attributed to inaccessibility, and such *extrinsic* isolation does not count as evidence of species-level divergence.

Use of this species concept has great value for comparing taxonomic diversity in widely different groups of sexually reproducing organisms, like orchids, ants, and ungulates. It can have indirect value in comparing diversity at higher levels. A genus that seems to be a natural group of ant species, recognized according to Mayr's criteria, should be roughly comparable to similar generic groupings of similarly defined species of orchids or ungulates. As many critics of Mayr's species concept have pointed out, it is only for a small proportion of the world's biota that we have any information on the presence or absence of intrinsic isolation between closely similar forms. For many groups, for instance those that never reproduce sexually, Mayr's criteria are not applicable even in principle. These practical difficulties do not detract from the usefulness of the two black basses as a suggestion on how different two organisms ought to be to put them in separate species.

Mayr's species concept gives us a standard level of taxonomic difference by which to recognize species distinction. The distinction corresponds to that of closely similar but consistently separate sympatric populations. Groups of closely related species, by Mayr's definition, could be recognized as a genus, the next higher taxonomic ranking. The species is thus a key taxonomic concept, but that is all it is. It is not fitting, in my opinion, to endow the species concept with theoretical significance beyond that of a widely recognizable level in the taxonomic hierarchy.

This frequent practice is a source of much mischief in biological discussions because: (1) some of the proposed properties of species are inconsistent with Mayr's or related taxonomic definitions (Sokal and Crovello 1970); (2) a lingering Aristotelian essentialism for the species category makes typological thought widespread, especially among physiologists, who often seem to expect that similar experiments on organisms with the same name should give the same results (examples in Ghiselin, 1974a); (3) ecologists often use the term *species* to mean 'that which occupies a niche in the trophic structure,' and this assumption is violated by the great majority of the world's species; (4) a species has been considered a historically unique *individual* (Ghiselin 1974b; Hull 1976), with the erroneous implication that taxa of higher or lower rank are merely parts or arbitrary collections of individuals; (5) morphologically defined species (e.g. of paleontologists) have levels of inclusiveness quite different from those studied in the field and laboratory, and their origin must often depend on changes that lag far behind or proceed far in advance of the sorts of speciation processes studied by neontologists; and (6) the species rank has been accorded unwarranted significance in clade selection by Gould and Eldredge (1977), Stanley (1979), and Vrba (1989).

Here I will not discuss the species concept in relation to the first two difficulties listed. I endorse many of the arguments made by Sokal and Crovello (1970), Ghiselin (1974*a*), and Holman (1987) on these matters. The discussion below relates to points three through six, on the species concept in relation to ecology, individuality, macroevolution, and clade selection.

8.5.1 Fallacy of species niche occupancy

Difficulties in the use of the species concept in community ecology are a matter that has lately received some attention, e.g. by Hengeveld (1988). It is now generally recognized that the niche of a single species may vary widely over its geographic range (James *et al.* 1984; Livingston 1988). As an extreme example, a mollusk that acts as a predator of a crustacean in one region is mainly that crustacean's prey in another (Barkai and McQuaid 1988). Effects of ontogeny on trophic relations can be extreme and have led some workers to speak of ecologically different developmental stages of a single population as separate *trophic* or *ecological species* (Briand and Cohen 1984), a practice criticized by Polis (1984). The changes can be decisive and qualitative at certain stages, as with insect metamorphosis or the sudden changes in the teeth and diet of certain cichlid fishes (Ribbink 1984). Gradual quantitative changes are no less important. Tuna of 1 g, 100 g, and 10 kg may be of the same population of the same species, but can hardly play the same roles in community ecology. An eloquent statement on the need for both theoretical and field work on the community effects of ontogenetic change in niche dimensions has been made by Werner and Gilliam (1984).

This whole difficulty would disappear if there were an acceptable substitute for *species* for community ecologists. Damuth (1985) suggests *avatar* for the collection of individuals of a single species found in a particular community. We still need terms for ecologically diverse developmental stages and morphs of the same avatar. A woodland may contain both tadpoles and adults of leopard frogs, or a stream have both planktivorous and benthic feeding morphs of the fish *Ilyodon furcidens* (Grudzien and Turner 1984). Such conspecific forms have ecological niches, most notably trophic positions, quite different from each other. I agree with Polis (1984) that it is misleading to call them *ecological species*, but neither he nor anyone else has proposed any substitute terminology. Polis does mention Livingston's (1982) use of *ontogenetic trophic units* (OTUs) for ecologically divergent year classes or other developmental categories, but some trophic units are neither taxa nor age groups. They could be seasonal visitors (Schneider 1978) or trophically specialized morphs of resident populations.

While Livingston's term may not serve all needs, it provides a basis for discussion. There could be ontogenetic trophic units (OTUs), taxonomic trophic units (TTUs), morphic trophic units (MTUs), or merely trophic units (TUs) for those concerned merely with a community's current trophic structure, and unconcerned with historical or other explanations for it. Unfortunately the priority of *OTU* (operational taxonomic unit) as a term in numerical taxonomy argues against a new meaning for ecologists. Perhaps developmental *trophic unit* (DTU) could be substituted.

There may also be a question on the need for assigning groups of organisms to TUs of any kind. Why not deal, initially at least, merely with individuals, in those communities for which individuality is meaningful (arthropods, vertebrates, many others). Each individual could, for example, be described as a point in a hyperspace of ecological measures. The clustering of many such points could then provide an objective basis for recognizing TUs. Such a system would certainly separate some trophic morphs and many ontogenetic stages, such as frogs and tadpoles.

8.5.2 Fallacy of species individuality

A clade is an individual in the codical domain, and a phylad in the material, in a particular sense. Every such entity has a more-or-less recognizable beginning, at least a presumed end, and a unique history and set of attributes in between. This is the sense in which Ghiselin (1974b) proposed that a species (e.g. *Homo sapiens*) is not a category with members but rather an individual with parts. This suggestion gained widespread acceptance, e.g. by Hull (1976, 1978). It was also criticized on many biological and philosophical grounds by Van Valen (1982), Ruse (1987), and O'Hara (1988). I wish here to emphasize two criticisms, the first being an absence of a decisive beginning for a species, an objection also raised by Bock (1986) and Wilkinson (1990).

Even the beginning of a new population or gene pool may be poorly defined. How absolute an obstacle to dispersal must a barrier be to make two populations allopatric? How far apart need they be to be considered isolated by distance? How strict must assortative mating be (in sympatric speciation) for a single population to be considered split into two? Species origins must be far more nebulous than population origins. Given that there are now two populations where there was only one, how much and for how long must they diverge before we can say that there are two species where there was only one? The answer can never be anything better than: when a neoDarwinian taxonomist thinks that the two forms will never again freely interbreed even if sympatric. This establishment of intrinsic reproductive isolation, as many studies have made clear, can be reached with widely variable levels of distinction in other biologically interesting properties, such as genetic distance, morphological contrast, and ecological divergence (Otte and Endler 1989).

It is clear that species origins, except for those originating by hybridization or other saltational event, will ordinarily be less decisive than population origins, and that a population may be much older than the species to which it belongs. Our own lineage may well be illustrative. In using molecular comparisons to decide among the possible cladistic relationships of *Pan*, *Gorilla*, and *Homo*, Smouse and Li (1987) found that the separation of the three lineages from a common ancestor must have been nearly simultaneous. They speculated that, for a while, the newly diverged chimpanzee, gorilla, and human lineages were only subspecifically differentiated members of the same species. This must have been followed, in our own lineage, by a time when *Australopithecus* and *Homo* were only subspecifically distinct, and then by a time when *H. sapiens* had diverged only slightly from *H. erectus*. As the known fossil record becomes more detailed, we can expect ever greater difficulty in defining the beginnings of any of these taxa. The recognition of individuality for a species or any other taxon is much less justified than for a population or monophyletic group of any taxonomic level of inclusiveness.

Another problem with the species-as-individual idea is its implication that clades of greater or lesser inclusiveness are not individuals in the sense claimed for species. If *Drosophila melanogaster* is an individual, can the same not be said for the culture I established on page 99. An important difference that can be recognized is that, if the taxonomists are right about *D. melanogaster* being a valid species, its end can come about only by extinction. My culture could also go extinct, but it might also end by fusion with another population. Clades of any level of inclusiveness can be shown by dendrograms, but those below the species level may show anastomoses in addition to the traditional branchings and extinctions (Fig. 2.1B, p. 14, and related discussion on pages 98–100).

The same considerations apply outside biology, for instance to institutional history, whenever new institutions arise by the fission of old ones. My academic Department, Ecology and Evolution at Stony Brook, is a unique individual that arose by the fission of an earlier Department of Biological Sciences. An effort to write the history of my present Department would hardly be inhibited or rendered illogical by the possibility of its ending by fusion with a currently independent administrative unit. Such a possibility likewise has no bearing on the possibility or fruitfulness of investigating the histories of subspecies and showing their genealogical relationships on dendrograms.

What might have been considered two species of Mesoamerican towhee may have fused in historical time (Bock 1986). The goose genus *Chen* provides a more current example. The blue goose and snow goose for many years seemed distinct species, although reproductive allopatry ruled out the actual use of Mayr's criterion of species-level separation. Now it appears that the distinctions are breaking down, with the two forms freely interbreeding in their region of contact on the west coast of Hudson Bay (Cooke 1988). A dendrogram of this and related genera would not be any less valuable for having to show that two branches fused after having independent gene pools for many years, or perhaps millennia. Like an academic department, the blue goose can be an individual for a while, and then end by fusion with another individual to form a new individual. It is now widely realized that both extrinsic and intrinsic isolation between populations may be temporary (Kat 1985; Eschelle and Connor 1989; Harrison and Rand 1989). The individuality concept as applied to clades and phylads of any rank is clearly a useful idea. Species individuality can only foster confusion. Concern for endangered species can be counterproductive when what is really needed is concern for *evolutionarily significant units* (Ryder 1986).

8.5.3 Fallacy of adaptive uniformity

Paleontologists, and systematists working with preserved specimens in museums, must mainly base taxonomic decisions on observed morphology and on analogy with previous work with similar specimens. Field notes and a general grasp of natural history will often be important aids. No one, for instance, is likely to base a species distinction on the difference between a caterpillar and a butterfly or between a seedling and a mature tree. The presence or absence of antlers on a mammalian skull is a major qualitative distinction and would be a most convenient key character. That antlers are not automatically used for taxonomic diagnosis shows that mammalogists recognize them as mostly diagnostic of age, sex, and season, rather than taxon, and realize that mammalian populations must contain both sexes.

Such essential biological information is often lacking for the interpretation of many characters, such as the sex-independent polymorphisms discussed in Chapter 7. A taxonomist knowing nothing of the ontogeny of the trophic specializations of the cichlid fish of Quatro Cienegas studied by Sage and Selander (1975), or those of Nicaragua studied by Meyer (1990), or of the goodeid *Ilyodon* studied by Grudzien and Turner (1984) would surely put them in different species. It used to be assumed that parasitic and nonparasitic life histories of lampreys provided major taxonomic distinctions. It is now known that both sorts of life cycle can occur in a single population (Beamish 1987). Genetically determined semelparity and iteroparity can be found in the same population of ascidians (Grosberg 1988). Likewise for presence or absence of paedomorphosis in a salamander (Semlich *et al.* 1990). Other examples of strikingly different morphologies and life histories were discussed in Chapter 7. It would appear that 'infraspecific macroevolution' (Liem and Kaufman 1984) is common.

The morphological and other sorts of diversification discussed above are found in single populations, and arise from genetic polymorphisms or environmental cueing of alternative ontogenies. If allopatric populations of a widespread species are considered, the range of variation is often far greater. It is a general rule that whenever what seems to be a widespread species is carefully studied it turns out to show great geographic variation and may include many species that meet the criterion of intrinsic reproductive isolation. There are now about 27 species of what used to be the leopard frog, *Rana pipiens* (Hillis 1988). It would appear from the results of a few studies in limited areas (e.g. Lavin and McPhail 1987; Baumgartner *et al.* 1988; Mori 1990; Foster 1992) that there may be thousands of species of three-spine stickleback.

Geographic variation conforms extremely well to the assumption of gradualism, as has been noted by many recent workers, e.g. Stenseth (1985), Slatkin (1987), Maynard Smith (1988*b*, p. 133). Adjacent populations are often closely similar, more geographically remote populations more divergent. This pattern and its implications for speciation are clearest in ring species, intrinsically isolated compatriots connected by a chain of allopatric intermediates. The classic example is the herring gull and blackback gull, which are separate species in the North Atlantic but are connected through Eurasia and the Pacific by subspecific intermediates. This and many similar examples are discussed by Mayr (1942, 1966). A recently studied example is a chain of montane salamanders distributed in a ring around the dry central valley of California (Wake *et al.* 1989). Gradualism in the spatial dimension can be shown in the fossil record (Brande 1979) and can only result from gradual change through time. Subspecies, whether they differ slightly or greatly, must have diverged from a common ancestor, and there are no clear distinctions among subspecies, semispecies, species complexes, and other degrees of relationship found by students of speciation (e.g. Darwin, 1859, Chapter 2; Otte and Endler 1989).

There should be little need to belabor the idea that two populations may be reproductively incompatible (i.e. two species by Mayr's criterion) without showing any marked morphological, behavioral, or ecological differences. Such populations have been called *sibling species* ever since reproductive incompatibility in culture led to the discovery of *Drosophila persimilis* as a species distinct from *D. pseudoobscura* (Futuyma 1986, p. 111).

Taxonomic problems arising from major within-species morphological differences and from absence of morphological distinctions between intrinsically isolated populations are extremely common, not isolated curiosities. The evolutionary independence of reproductive incompatibility from genetic, morphological, ecological, and behavioral divergence has been recognized and stressed by many recent workers: Raven (1986) and Templeton (1989) for plants, Larson (1984) for salamanders, Chow *et al.* (1988) for crustaceans, and many workers (Shapiro 1984; West-Eberhard 1986; Turner 1988) for insects. The study of evolutionary innovation and adaptive radiation must be conceptually decoupled from that of the origin of intrinsic reproductive isolation, or the specific mate recognition systems discussed by Lambert and Paterson (1982). Cladogenesis is an essential concept, but speciation in the usual sense has no special significance for macroevolution. This is a conclusion reached by many recent workers, e.g. Shapiro (1984), West-Eberhard (1987), Otte (1989), Lessios and Cunningham (1990).

8.5.4 Fallacy of species selection

There is no reason why *species selection* (Stanley 1979; Vrba 1989) should be recognized as a special process different from any other kind of clade selection. Wynne-Edwards' (1962) original concept of group selection dealt with selection among clades that were not intrinsically isolated, and the bulk of subsequent group-selection modeling (Maynard Smith 1976; Wade 1978) conforms to this tradition. Selection can likewise take place among clades of higher than the species level of taxonomic diversity, as argued by Van Valen (1975, 1988). Van Valen and Maiorana (1985) proposed a possible example, the competitive replacement of brachiopods by pelecypods during the Mesozoic. Vermeij (1987, pp. 85–9) thinks this may be attributable to the pelecypods' superior feeding apparatus.

The picture that emerges is utterly at variance with the theory of punctuated equilibria derived by Eldredge and Gould (1972, 1988) from the fossil record and the peripheral isolate theory of speciation (Mayr 1954; Carson 1989). They find that species commonly change but little in millions of years, and that morphological changes, when they occur, are sudden and associated with speciation. They propose that new species arise, not by gradual divergence of major divisions of widespread species, but by rapid divergence of peripheral isolates, small populations near the edge of the range of the parent species. Not all paleontologists agree that the fossil record usually shows the pattern described by Eldredge and Gould (Gingerich 1977; Levinton 1986, 1987), but I will assume here that there is some factual validity to the pattern they describe.

There are two possible resolutions to the conflict between the punctuated equilibrium interpretation of the fossil record and the gradualist pattern of contemporary variation and biogeography. Either speciation and other evolutionary processes are quite different in all major groups today from what they were from the Cambrian to late Pleistocene, or the appearance of punctuated equilibria in the fossil record is misleading. Ever since Darwin (1859, Chapter 10), paleontologists have recognized and discussed the imperfections of the fossil record. I leave it to modern specialists to decide whether this record can be interpreted as documenting the gradualism through time that must underlie the Recent biota's gradualism through space. If not, the contradiction is indeed serious.

The peripheral-isolate theory of speciation is another matter. It was proposed as a way of producing the genetic revolution thought to be needed for evolving the new adaptations required for occupation of a new ecological niche. Since we now know, on the basis of the works discussed in this and the previous chapter, that radically different morphologies and ecological niches can be found within species, and even within single populations, the peripheral-isolate theory would seem to be of little use. Recent theoretical work suggests that small peripheral populations would not be especially adept at rapid evolution into new niches (Rouhani and Barton 1987; Charlesworth and Rouhani 1988; Barton 1989).

My reservations on punctuated equilibria as descriptive of evolutionary history, and rejection of the peripheral-isolate theory of speciation, does not imply any doubts on the reality of prolonged stasis in the fossil record, a problem that deserves at least its own chapter.

9

Stasis

Critics of the theory of natural selection since its first announcement have attacked it with a time-shortage argument. How could one really believe that accidental variations could combine to produce anything as complex as an amoeba, let alone a human being, in even the most liberal estimate of the duration of the Earth's history? The critics offer such analogies as: how long would you have to wait for a monkey playing with a typewriter to produce a perfect copy of Hamlet's soliloquy? They thus show how totally they have failed to understand the logical structure of the theory. They fail to see the elementary distinction between evolution by the survival of the fittest, and spontaneous generation followed by survival or extinction. This folly misleads not only the untutored, but can be found in the publications of supposedly expert scholars. Waddington (1953) provides one example, the book *Mathematical Challenges to the neo-Darwinian Interpretation of Evolution* (Moorhead and Kaplan 1985) several others.

I am sure I could get a monkey (or preferably a logistically less demanding but equally illiterate computer) to compose Hamlet's soliloquy in a few minutes. I need only impose a program of artificial selection that would preserve everything the computer-monkey did that made its cumulative effort resemble Hamlet's words more closely, and reject all changes that decreased the resemblance. The idea of producing some specified text by the editing of randomly placed symbols may now be an obvious analogy to the Darwinian process (Ruse 1982, p. 308; Dawkins 1989). I first heard it suggested by F. E. Warburton about 1965.

But correcting the main misunderstanding does not get rid of the question of the adequacy of geological time. Is a few billion years enough for the evolutionary process to produce the complexity and diversity of life that we now find? I believe that the intuitive answer to this question, for many educated people, is *no*. Mammals and insects and trees are so endlessly complicated, from the grossest down to the most microscopic structures, and so strikingly different from each other. Surely it is unreasonable to claim that so slow and unguided a process as the environmental testing of rare mutations would be capable of such accomplishments. In response to this kind of thinking, some biologists have sought for ways in which the evolutionary process might be speeded up or made more immediately responsive to the needs of the organisms. I offer Goldschmidt's (1940) mutationism and Waddington's (1953) genetic assimilation as examples of such thinking. A few biologists all through this century have gone further and rejected Darwinism in favor of some sort of Lamarkism.

The question of evolutionary rate in relation to available geological time is indeed a serious theoretical challenge, but the reason is exactly the opposite of that inspired by most people's intuitions. Organisms in general have not done nearly as much evolving as we should reasonably expect. Long-term rates of change, even in lineages of unusually rapid evolution, are almost always far slower than they theoretically could be. The basis for such expectation is to be found most clearly in observed rates of evolution under artificial selection, along with the often high rates of change in environmental conditions that must imply rapid change in the intensity and direction of selection in nature.

9.1 Taxonomic stasis

While artificial selection no doubt provides the clearest evidence on the rates of evolution of which organisms are capable, it is not necessary to invoke experimental or agricultural data for this argument. Short-term evolutionary rates in nature provide abundant evidence without the stigma of unnaturalness. Endler (1986) documented rapid responses to selection in nature for a large number of diverse organisms. Many other examples have been described in the few years since Endler's review. Morton (1990) found evidence that the hooded warbler has markedly modified plumage pattern and habitat selection since the invention of agriculture. Andren *et al.* (1989) showed that a frog has evolved much greater tolerance of acidification in less than 40 years (*ca.* 15 generations). Hairston (1990) found that freshwater copepods rapidly evolve defenses against introduced predators. Billington *et al.* (1988) showed that plant populations in two adjacent fields diverged noticeably in gene frequencies as a result of one field being fertilized regularly and the other not. Houde (1988) found noticeable changes in color from a few generations of divergent sexual selection in guppy populations. Various guppy behavior patterns also respond rapidly to environmental change in nature (Stoner and Breeden 1988). Changes of about 5% per generation (several generations per year) occurred in mosquitofish introduced into new habitats (Stearns 1983, not cited by Endler 1986). The genetic basis for the changes in the guppies and mosquitofish were demonstrated in these studies, just a few of what could be cited from just this one family of fishes.

A less extreme but long-term and thoroughly quantified example may be seen in comparisons among introduced North American populations of English sparrow and with their British source. Parkin (1987) notes that variation in North American populations has begun to reconstitute, in some respects, the Old-World divergence into four subspecies. Parkin's study deals with both molecular and morphological changes during the century and a half of English sparrow evolution in North America. In their classical morphological comparisons, Johnston and Selander (1964, 1971, 1973) found that certain linear measures, such as lengths of various bones, had changed by as much as 5% in about a century. At this rate, no birdwatchers will notice in their old age that the bird looks any different from what they remember from childhood. From most people's perspectives, English sparrows are staying much the same.

A bit of arithmetic gives another kind of perspective. If tibiotarsus length had been increasing at 0.05 per century throughout the Pleistocene (traditionally 10,000 centuries), the length now would be greater than the original by a factor of e^{500}, a number my calculator refuses to deal with. A more manageable approach is to consider how long it would take for the tibiotarsus to go from a centimeter to a meter in length. My now cooperative calculator tells me it would take about 92 centuries, less than 1% of the Pleistocene. The observed rate of evolution would be capable of turning sparrow-size into ostrich-size bones, and back again, about 54 times during this geologically trivial period.

Many explanations have been offered for the discrepancy between what we might expect and what seems to happen. Strong directional selection in the laboratory often shows marked effects for only a few generations. Additive genetic variation may be quickly exhausted, and further selection can have little effect. We might imagine that similar limitations could overtake Houde's guppies and Johnston and Selander's sparrows, with perhaps little further change of the kinds observed, despite continued selection.

It would be unwise to rule out this kind of explanation for some examples of stasis in nature, but current understanding indicates that it is unlikely to be generally applicable. Response-to-selection experiments usually make use of limited numbers of individuals that could provide only a small amount of the genetic diversity found in wild populations. More important is the fact that artificial directional selection is normally far stronger than is likely to be consistently maintained in nature. Weak selection would allow for the regeneration of genetic variability through mutation, gene flow, and other processes. Ayala's (1982) calculations show that even quite conservative estimates of mutation rates can be sufficient to prevent any serious shortage of genetic variability in quantitative characters in nature. It should also be noted that domesticated animals and plants of a wide variety of taxonomic groups can still show marked responses to directional selection, despite centuries or millennia of what, by natural standards, were extraordinarily rapid rates of evolution (Crow 1988).

Against this it can be argued (Eldredge 1989; Vermeij 1987, p. 5; Wake *et al.* 1983) that great climatic and other changes in the physical environment during the Pleistocene need not mean that the environments experienced by organisms are changed to any important degree. A characteristic group of organisms is adapted to the tundra, and they are now found in the North American and Eurasian arctic. Fifteen thousand years ago they may have inhabited tundra in places like New York and Bohemia. As their environment shifted north, the organisms shifted north (or up into the Alps) and continued to experience the same selection pressures from similar environments.

Unfortunately this fable will not bear close examination. The physical environments of arctic and alpine tundras differ in many ways from each other and from the mid-latitude, low-altitude tundras of the Pleistocene. They differ in seasonal and diel photoperiod and seasonal amplitude of insolation, in magnitude and seasonal distribution of snowfall and rainfall and associated soil moisture, in partial pressures of oxygen and other gases. Even these few environmental differences imply different selection pressures now from those that prevailed 15,000 years ago.

The physical changes are probably of minor importance compared to biological differences in community composition. The shifting climatic zones of recent millennia were not followed by shifting distributions of intact communities (M. B. Davis 1989). Different species responded to changes in different ways. Some expanded their abundances or ranges while others grew sparser, or became relics of reduced range, or went extinct (Pease *et al.* 1989). Late Pleistocene climatic shifts may have been faster than some species could move into newly favorable regions. This was probably true for trees such as spruce and beech that can be dominant elements in many terrestrial communities (Roberts 1989; Gear and Huntley 1991). Currently important tree species in Ontario may have arrived a millennium or more apart (Liu 1990). It seems unlikely that the community composition of a particular time and place could be closely matched anywhere after a major climatic shift. Any population of a continental habitat today must be dealing with a rather different array of resources, competitors, predators, and parasites from those of the most recent Pleistocene temperature minimum.

The Pleistocene included several major climatic revolutions and associated changes of sea level and alternating bouts of isolation and mixing of biotas. Tropical communities were affected by these biotic changes and by shifting geographic and seasonal distributions of rainfall. The general level of evolutionary stasis is difficult to reconcile with this great environmental flux. The current facts and understandings of population genetics would be thoroughly compatible with major changes in the adaptations of most lineages of animals and plants during the last million years.

Instead, a large proportion of Recent species are essentially identical to their Pliocene ancestors and, of course, far more impressive examples of taxon stasis can be found. Carroll (1988) provides many examples from the vertebrates. There are fish skeletons from Cretaceous deposits that are nearly the same as can be found today (Fig. 9.1), and many Recent genera may be identified from Cretaceous deposits. Usually such fossils are given their own species names, but this nomenclatural practice is not always justified by demonstrated differences that would rule out inclusion

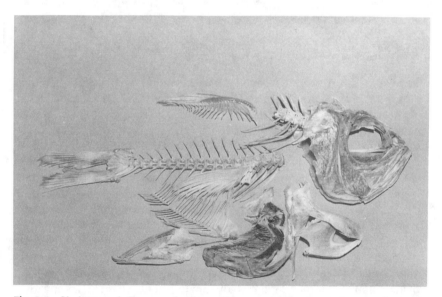

Fig. 9.1. Skeleton of *Beryx splendens*. Congeneric forms are known from the Cretaceous, when they lived in communities with ichthyosaurs and ammonites. The skull, which appears here as a single structure, is really a complex of more than a hundred elaborately articulated pieces. The detailed structural similarity between Recent and Cretaceous forms implies nearly identical locomotor and trophic adaptations and ways of life for about a hundred million years. More extreme examples of stasis could be listed. Photograph by the author, of specimen 095739SD, in the American Museum of Natural History.

of a Cretaceous form in a Recent species. I find it puzzling that no shift in selection pressures in the last 100 million years would have produced any noteworthy change in a structure as complex and informative as a fish skeleton, which ought to reflect even very subtle changes in locomotor or trophic adaptations.

Even some widely recognized examples of rapid evolution are really extremely slow. Data on Pleistocene human evolution are interpretable in various ways, but it is possible that the cerebrum doubled in size in as little as 100,000 years, or perhaps 3000 generations (Rightmire 1985). This, according to Whiten and Byrne (1988) is 'a unique and staggering acceleration in brain size.' How rapid a change was it really? Even with conservative assumptions on coefficient of variation (e.g. 10%) and heritability (30%) in this character, it would take only rather weak selection ($s = 0.03$) to give a 1% change in a generation. This would permit a doubling in 70 generations. An early hominid brain could have increased to the modern size, and back again, about 21 times while the actual evolution took place. Indeed, it is plausible that a random walk of 1% increases and decreases could double a quantitative character in less than 3000 generations.

9.2 A desperation hypothesis

Whenever we look for it we find rapid evolution in the Recent biota. Simpson's (1944) horotelic rates (far greater than normally found in the fossil record) seem the general rule. Likewise in species and subspecies taxonomy and biogeography we consistently find patterns indicative of horotely during recent millennia. It is clearly not true that 'species have . . . genetic and developmental coherences that resist selective pressures of the moment' (S. J. Gould 1983, p. 362). How can these observations of rapid evolution be reconciled with the stasis often (usually?) found in the fossil record for periods of millions or tens of millions of years? In Chapter 8, I took the traditional course of blaming the imperfections of the fossil record, but what kind of imperfection must be postulated to explain the appearance of widespread stasis in a biota in which most populations are evolving rapidly? I suggest that the record simply fails to provide enough short-term detail on phylogenetic changes at any one time, such as those in progress in the complexes of species referred to as leopard frogs (Hillis 1988) or threespine sticklebacks. Perhaps it thereby fails to document what might be called *normalizing clade selection*.

The threespine stickleback is a cosmopolitan coastal fish of cool-temperate and boreal climes of both sides of the North Atlantic and Pacific Oceans. Often along these coasts it invades confluent fresh waters where it may specialize for local ecological conditions. Canadian examples

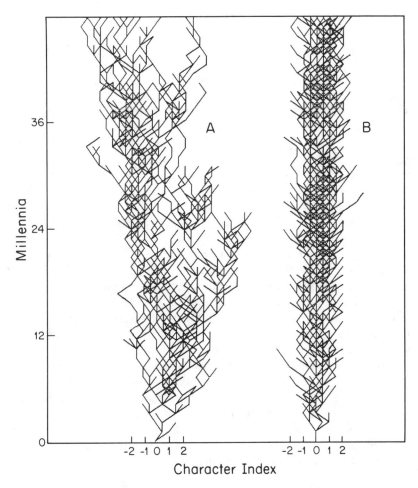

Fig. 9.2. Dendrograms generated by random branching and character changes. The x-axis shows population means of a metrical character measured in standard deviations, with the starting population mean set at zero. After each millennium of its history (y-axis) each population may change its mean value by -1.0, -0.5, 0.0, 0.5, or 1.0 standard deviation, with relative probabilities 1,4,6,4,1. After each millennium each population in A is represented by a number of descendant populations equal to a random value from a Poisson distribution with $\lambda = 1.0$. After each millennium each population in B is represented by a number of descendant populations equal to a random number from a Poisson distribution with λ determined by x. The mean expectation is 1.2 with the ancestral $x = 0$, but it changes by decrements proportional to the square of the deviation from 0. It decreases to 1.0 at $x = 1$ or -1. Thus A shows a completely random phylogeny while B shows random changes opposed by normalizing clade selection. A few fossil specimens from B could easily be read as a record of stasis arising from internal constraints on evolutionary change. In reality the changes are generated

133

are presented by Bentzen *et al.* (1984), Foster (1992), and Francis *et al.* (1986), Asiatic by Mori (1990) and Zinganov *et al.* (1987). If every population with ecologically significant distinctions were accorded formal subspecific recognition, and every intrinsically isolated form a species name, the taxonomic complexity would overwhelm a faunal work on any region in which this species group occurs.

Conceivably we are witnessing the start of a major adaptive radiation of freshwater fishes from a single marine ancestor, but I prefer the alternative suggested by M A Bell (1989). The phylogeny of the threespine stickleback is not so much a tree as a raceme (Fig. 9.2). It consists of a single geologically persistent species that frequently gives rise to temporary freshwater forms. Clade selection acts against freshwater populations, either because they can not compete in mature freshwater faunas or because their habitats and ecological niches are ephemeral. The freshwater forms come and go in rapid succession, but the species complex endures in much the same form for long periods of time. Bell and other workers (cited above) have amply demonstrated the rapid speciation of threespine sticklebacks in habitats that were not there a few thousand years ago. Not surprisingly, there are no data as yet on the implied rapid extinction and intense clade selection against all but the conservative marine form.

This pattern of evolution may be unusual, or perhaps the threespine stickleback is merely an unusually clear example of a common pattern. A diagram of punctuated equilibria (e.g. Fig. 2.1C) could be made to illustrate the sort of racemose phylogeny proposed for sticklebacks by making each of the lineages complex and bristly, like Fig. 9.2B. Most species would be constantly breaking up into divergent lines specializing for new habitats or ways of life. Then after a few millennia most such lines have gone extinct. The remaining few nearly always include a form close to the ancestral type. If this form were numerically dominant and

---

Fig. 9.2. Continued
with great facility but are opposed by clade selection consistently favoring the ancestral way of life. If apparent stasis is often as illustrated in B, Fig. 2.1D would be more realistic if its nearly vertical lines were replaced by complex columns of branchings and extinctions as shown in B.

The great majority (>90%) of both kinds of simulation show extinction long before the 48,000 years shown. A was the third most phenotypically diverse phylad of the first five that lasted the full 48,000 years. B was the phylad with the third largest number of descendant populations of the first five that lasted to the end of the run. Long-lasting simulations of random phylads (A) are quite diverse, while those subject to normalizing clade selection (B) are, understandably, much more uniform.

most widespread, like the marine stickleback, it would be most likely to contribute to the fossil record and appearance of stasis.

The usual persistence of the ancestral type would follow from purely statistical considerations. Not all habitats or ways of life need be equally durable. If we had to predict which populations will still be represented by descendants a million years from now, we would do well to choose those that have already persisted with little change for millions of years, rather than their recent offshoots of mere thousands or tens of thousands of years in age. The appearance of stasis in the fossil record would result from an enormous variability in the persistence of ecological niches.

How common are racemose phylogenies (Fig. 9.2B)? Any answer at the moment would be pure guesswork, but I hope that M. A. Bell's (1989) work will alert others to similar possibilities. It may be that the loss of sexuality in evolving lineages is far more common than suspected but consistently opposed by clade selection so that the great majority of eukaryote organisms show at least occasional sexual reproduction (Chapter 10). It may also be that even minor restrictions on recombination, such as would result from loss of floral heterostyly (Barrett 1989; Barrett *et al.* 1989), are frequently evolved but consistently lead to extinction. Salamanders with terrestrial adults and aquatic larvae often evolve into paedomorphic forms that reproduce as sexually mature larvae. Populations with this simplified life history have a high rate of extinction, but are readily replenished by offshoots from the more persistent populations with the ancestral life cycle (Schaffer and Breden 1989). Island lizards often evolve a small body size but then go extinct when the island is invaded by the larger ancestral form (Roughgarden and Pacala 1989). Wherever the evolution of a population is irreversible, or even goes in one direction more readily than the other, the persistence of the less easily evolved condition demands favorable clade selection.

There is a clear analogy at the genic level. If mutation goes one way more readily than the reverse, the persistence of the allele less favored by mutation pressure shows that a higher-level process, the natural selection of alleles at that locus, must be operative.

9.3 Character stasis

Even in groups that evolve rather rapidly through geologic time, the Hominidae being a good example, with most lineages changing their taxonomic status in a million years or less, there can be a striking stasis of particular character states. The rest of this chapter will consider some outstanding examples of characters that, I would think, ought to be labile in evolution and tailored to ecological circumstances, but instead show extreme stasis. There may be those who think my expectations inappropriate. R. Levins (1968, p. 6) proposed that evolutionary biologists 'are more interested in qualitative than quantitative results,' and that their 'objective is not so much the discovery of universals as the accounting for differences.' I presume that, to Levins at the time, the absence of a difference was the absence of a problem. My goals are more ambitious. I would like to know the circumstances under which certain patterns and quantitative levels of variation among clades will always be found. I think Vermeij (1987, p. 26) had something similar in mind when he said 'we wish to know why certain types of organism do not exist.'

I argued in Chapter 5 that quantitative parameters of evolved machinery are expected to respond to directional selection in Mendelian populations. The expectation is often but not always realized. I think evolutionary biologists should strive to replace this 'often but not always' with a simple and decisive sort of 'always if. . .' and thereby achieve the sort of universal on which Levins was pessimistic. On a geological time scale the sort of response to selection I have in mind should be so rapid that populations would normally be near their evolutionary equilibria. There should be no appreciable difference between mean and optimum, for characters subject to optimization (Chapters 5 and 6). In Maynard Smith's (1978) terminology, the *lag load* should be negligible. This assumption is implicit in any use of ESS or optimization models in predicting the outcomes of investigations. The sorts of exceptions to the expectation of optimized character values that I discussed in Chapters 5 and 6 are easily accommodated by current evolutionary theory.

9.4 Avian and mammalian body temperature

Other exceptions are not readily accommodated. The set points of body temperatures of the two groups of endotherms are a good example. Temperature is very much a continuous variable. Yet human body temperature seldom deviates more than a degree from 37°C, and this is a typically mammalian value (Prosser 1973, pp. 394–412). Those mammals traditionally considered primitive (marsupials, monotremes, edentates) may have lower and more variable temperatures (as low as 31°C in active specimens) but most eutherians are remarkably uniform. Prosser gives either 38.0° or 38.1° for camels, baboons, and fur seals. Birds are similarly conservative, generally keeping their body temperatures in the 38° to 42°C range.

The data make no evolutionary sense. Why has there not been a major adaptive radiation of temperature set points? Would there not be a great energy saving for a snow bunting or arctic fox that kept its body temperature at 28° rather than 38°? Would it not be advantageous for a gazelle or baboon fleeing from hyenas over a sun-baked Arabian plain to operate at 48°, and likewise for the hyenas? Why should there be such enormous diversity in a long list of characters in both birds and mammals (think of whales, anteaters, bats, penguins, hummingbirds, moas) but so little in this one character?

An undoubtedly correct but unsatisfactory answer to all these questions is that some sort of constraint prevents any appreciable evolution of body temperature in these organisms. The answer is unsatisfactory until we can describe the nature of the constraint or the nature of the difficulties that would arise from any incipient evolution of a higher or lower set point. The constraint is relaxed in special ways. There are bats and rodents and small birds that can have torpor temperatures down to 5°C (or even below 0°C (Barnes 1989)) and then spontaneously warm up to the normal active range with no damage from the chill. Our own peripheral tissues can be cooled many degrees below core temperature without serious impairment.

McArthur and Clark (1988) propose that the set points of mammalian and avian temperatures represent an ideal compromise between such needs as heat dissipation in hot environments, heat retention in cold, and the avoidance of dessication. I am sure that these are factors in the evolution of the set points, and McArthur and Clark make it clear that the range of set points should be much less than the range of habitat temperatures. They do not make it clear that the set points should not be subject to any adaptive differences at all between widely different animals. That elephants and lemmings should keep their bodies at very nearly the same temperature is mightily mysterious.

It may be that some artificial selection experiments could provide insights. I wish someone with a large mouse colony would try selecting for high and low body temperatures and let me know the results after perhaps ten generations. Individual mice do differ consistently in temperature by as much as a degree, and the variation has a heritability of about 0.11 (C. B. Lynch *et al.* 1988). No matter what the results of this selection experiment, they would be of enormous interest.

The temperature problem is compounded in mammals by the constraint on spermatogenesis. Why must this one process, among all the others in the life history and physiology of this class of vertebrates, require a temperature about a degree lower than the others? Meeting this requirement is surely costly. It requires special morphogenetic machinery to produce the special structures and circulation patterns needed to provide the reduced temperature. Another cost is the undoubtedly greater vulnerability to injury for testicles housed in a thin-walled scrotum. The seasonal testicular movements of some mammals may be considered a switch between protective housing, when spermatogenesis is unnecessary, and the more vulnerable scrotal position when the cost of mechanical hazards is worth the fertility benefits of reduced temperature. Grasse (1955, p. 467) summarizes what is known of the diversity of adult testicle positions in the Mammalia.

Why is it not a simple matter to achieve a minor shift in the temperature optimum for spermatogenesis? The fact that only a few mammals have been able to do so is surely an important fact that should be getting close attention from all biologists interested in both evolution and in mammals.

9.5 Electrolyte concentrations of marine vertebrates

There is an analogous problem for tissue electrolyte concentrations of marine vertebrates. This character may not be quite so phylogenetically inert as the body temperatures of either birds or mammals, but the stasis prevails over a far greater range of phylogenetic diversity. Marine organisms in general, both plants and animals, are osmoconformists. The osmotic properties of their tissues closely match those of seawater, despite some major differences in ion ratios. Primitive marine chordates, including the hagfishes, are similar to invertebrates in this respect. All other marine vertebrates are exceptional in having electrolyte concentrations far lower than that of seawater. The traditional historical explanation (Romer and Grove 1935) is that all vertebrates above the hagfish derive from a freshwater ancestor, for which reduced concentrations were clearly advantageous.

Whether this is or is not the correct historical explanation, the prevalence of low electrolyte concentrations in marine vertebrates results in special hazards, and its maintenance requires expensive mechanisms. Exposure of gills and other permeable surfaces to seawater always results in withdrawal of water from the blood and an increased concentration of solutes. This process is countered in bony fishes by special modifications of gills and guts and kidneys that provide for the energy-expensive work of excreting salt and moving water against an osmotic gradient. The cartilaginous fishes solve the problem with osmoconformity, not by allowing ion concentrations to rise to levels close to those of seawater, but by having a high concentration of urea in the tissues. If it were not for sharks and their relatives, urea would be considered a universal toxin, less lethal than ammonia, but tolerable only in small traces in any animal. Sharks have evolved a tolerance to a urea concentration of about 1%, far higher than that of other vertebrates. Why was this easier to achieve than an increased tolerance of common ions, the normal condition in hagfish and all marine invertebrates?

There is another disadvantage in low solute concentrations for marine fishes of middle and high latitudes. Their tissues freeze at warmer than −1°C (Prosser (1986, pp. 305–6) gives −0.7° as a typical value) while the water they are in remains liquid to about −2°. A fish in seawater at −1.5° may be supercooled and vulnerable to mechanical destruction from the formation of ice crystals in its tissues. Fishes in habitats subject to freezing may increase their ion concentrations a bit, but they mainly rely on higher levels of a variety of organic substances, analogous to the elasmobranchs' use of urea (Knight and DeVries 1989). An arctic fish may thereby have a freezing point of −2.7°C (Prosser (1986, pp. 305–6).

The reduced electrolyte levels evolved by freshwater fishes are demanded by all their descendants in marine environments. The tearful eyes of a sea turtle laying eggs on a beach result, not from any emotional impact of motherhood, but from her use of lachrymal glands to secrete a concentrated brine to keep her internal electrolytes at the usual vertebrate level. The runny noses of the albatross and other marine birds have the same significance. Marine tetrapods generally parallel desert inhabitants in their evolution of mechanisms for water retention. This is ironic in an environment that is more than 96% water and inhabited by hosts of other organisms for whom water is present in excess. I wish someone with a large colony of stenohaline mollies or other short-generation marine vertebrate would try selecting for increased tissue electrolyte concentrations and let me know what happens.

I do not doubt that some physiological reasons can be found for the low set point of total ion concentrations. Experimental alteration of salt balance of some fish or seal or sea turtle would no doubt cause demonstrable impairments of some enzyme functions. This merely shifts the formulation of the problem from the animal to its enzymes. Similarly I suppose that some specific biochemical processes can be identified that are seriously impaired by a few degrees of fever or hypothermia in a mammal, but why are mammalian enzymes so extraordinarily sensitive to temperature change, and why must all mammals demand almost the same temperature?

9.6 Why no viviparous birds or turtles?

The universal retention of oviparity in birds is a topic of much recent discussion (Blackburn and Evans 1986; Anderson *et al.* 1987; Dunbrack and Ramsay 1989). Blackburn and Evans review the relevant facts and previous explanations for the absence of viviparity in birds. Birds date from the Jurassic and are now represented by thousands of species of diverse structure, size, habitat, and way of life. Viviparity has been evolved more than a hundred times among vertebrates, most of these among the squamate reptiles. Viviparity ought to offer enormous advantages for some birds. The nesting period is the most dangerous in the life history of both young and parents, and the need for egg incubation greatly lengthens this period. Much of the danger would be avoided if the female would simply retain the eggs until hatching.

Many of the explanations offered attempt to show that certain features of avian organization would be incompatible with viviparity. Flight would in some way rule it out, or physiological requirements of the embryo could not be met within the body of the mother, or the avian immune system would not tolerate anything resembling placental contact between maternal and embryonic tissues. Blackburn and Evans (1986) show that these proposals are all untenable: there are many groups of flightless birds; many diverse groups of other vertebrates have overcome the physiological and immunological problems. Blackburn and Evans favor a denial of the supposed advantages: '. . . the costs of egg retention associated with decreased fecundity, increased maternal mortality, and decreased paternal investment outweigh the potential benefits for most birds.'

Anderson *et al.* (1987) and Dunbrack and Ramsay (1989) reject this conclusion mainly for its lack of broad applicability. Many bird groups have evolved the theoretical minimum of clutch size (one) and paternal investment (zero). For marine birds, which generally have small clutches, the necessity for terrestrial incubation would seem to be far worse than any disadvantage of pregnancy for the females. These authors propose instead to shift the problem of the universality of birds' oviparity to the universality of their temperature constraints, especially in relation to the great oxygen demand by an initially large avian embryo. They suggest that the bodies of adult female birds are too hot for rapidly developing young. Nowhere, supposedly, in the ancestry of all the thousands of Recent bird species has any lineage been able either to lower the maternal temperature during incubation nor to raise the embryonic temperature tolerance a degree or two. Both Blackburn and Evans (1986) and Anderson *et al.* (1987) suggest an experimental approach to the problem of incubation temperature. Eggs could be artificially incubated at the usual 40°C adult body temperature for varying periods after oviposition to determine the threshold duration that would cause observable damage. Artificial selection for extended tolerance could be carried out in a suitable species. Any appreciable success for such an endeavor would count against embryonic temperature requirements as an explanation for the failure of any bird to evolve viviparity.

A related and perhaps more serious problem has not, to my knowledge, had any comparable attention. Why are there no viviparous turtles? Here we can hardly invoke factors such as temperature constraints either directly or in relation to oxygen demand (Dunbrack and Ramsay 1989). Webb and Cooper-Preston (1989) suggest that turtle and crocodile eggs are specialized for early independence in a way that would make viviparity difficult to evolve. The less specialized respiratory adaptations of lizard and snake eggs would be less constraining. There is no other obvious impediment to egg retention in turtles that would not apply to snakes and lizards, which have evolved viviparity many times (Shine 1985; Shine and Guillette 1988). Oviparity seems especially maladaptive for sea turtles. Eggs buried on beaches are readily exploited by terrestrial predators. The emerging young must scramble across the beach to the sea and are taken in large numbers by predaceous birds. These hazards would be avoided completely if the young were born at sea. Another burden is imposed on the female when she comes ashore to lay. It is in these egg-laying trips that she is vulnerable to human and perhaps other large predators. It is only for this one part of the life history that ability to locomote on land is required. Copulation takes place in the water and males need never come ashore.

The theory of natural selection seems to demand that the first steps toward viviparity be taken by any turtle population in which a briefer period of development on the beach would result in reduced loss of the young, without some completely compensating disadvantage to the female in carrying the embryos for a longer time. The result would be decreasing periods of vulnerability for the eggs for as long as benefits exceeded costs. If conditions hold for an appreciable period of evolutionary time, viviparity would result. A viviparous sea turtle would be freed from all need to come ashore and we might imagine it to have a clade-selection advantage that would make it the ancestor of a diverse group of exclusively viviparous turtles.

It is easy to imagine that the preadaptations for the evolution of internal fertilization would often be lacking, and it would be difficult to evolve internal development without internal fertilization. In general the fishes conform to expectations here: the evolution of internal fertilization is usually followed by the evolution of viviparity (Breder and Rosen 1966). The universality of internal fertilization in the squamates does not require the evolution of viviparity, but it obviously permits it. The universality of internal fertilization with total absence of viviparity in the turtles and birds seems most mysterious. Viviparous sea turtles provide an outstanding example of the kind of non-existence that Vermeij (1987) and I would like to explain.

9.7 Other problems of character stasis

Meristic characters sometimes show remarkable stasis. In Chapter 1 I mentioned the near universality of seven cervical vertebrae among the mammals and suggested that each vertebra and associated nerves and blood vessels were so specialized that any loss or gain of a vertebra would be functionally disruptive. Analogous reasoning has been used to explain the stasis of low clutch or litter sizes (one or perhaps two) in some groups of organisms (Ashmole 1971; Anderson 1990). A similar explanation may be valid for many low-value meristic characters, such as the three anal fin spines of the great majority of percoid fishes. But what about vertebral counts and caudal elements in this group? Most of the thousands of species in dozens of families have 17 caudal fin rays (Grasse 1958). And what about the leeches' constant count of 34 body segments? Thirty- and 40-segment leeches are among the large numbers of organisms for which non-existence is urgently in need of explanation. Many other examples of unexplained character stasis can undoubtedly be identified in various taxonomic groups.

10

Other challenges and anomalies

In my view the various examples of stasis, some discussed in Chapter 9, form the most serious set of difficulties facing evolutionary theory today. In this chapter I try to continue in the spirit of Darwin's Chapters 6 and 7, and my own Chapters 8 and 9, and discuss various other theoretical challenges. Other investigators can undoubtedly identify major anomalies that I have overlooked. To these might be added an enormous list of single investigations that yielded results somewhat different from what had been expected. Some of these might well be of enormous importance, given the proper interpretation. The finding of mitotic irregularities and unexpected pigment patterns in maize leaves and kernels might have been shrugged off by many geneticists as a minor technical annoyance. Happily McClintock (1965, 1987) did more than shrug. She took the trouble to keep detailed records of the annoyances through successive generations and thereby discovered the previously unsuspected transposable elements.

10.1 Haldane's dilemma

Twenty years ago this was a much debated matter. Discussions began with J. B. S. Haldane's (1937, 1957) calling attention to the seemingly great demographic cost of natural selection, gained greater attention with the growing recognition of high levels of genetic variability in nature (history reviewed by Ayala (1982)), reached a high prominence with a decisive demonstration of this phenomenon (Lewontin and Hubby 1966), and climaxed in Bruce Wallace's (1970) book *Genetic Load* (updated by Wallace (1987, 1989) and Reeve *et al.* 1988). A formally rather similar problem is currently worrying investigators of the evolution of developmental mechanisms (Kauffman 1987; Wimsatt and Schank 1988).

In my opinion the problem was never solved, by Wallace or anyone else. It merely faded away, because people got interested in other things.

They must have assumed that the true resolution lay somewhere in the welter of suggestions made by one or more of the distinguished population geneticists who had participated in the discussion. Those who were trying to keep abreast of evolutionary issues around 1970 will be thoroughly familiar with this episode in the history of evolutionary thought. For those who were too young for such pursuits in 1970 I offer the following sketch of Haldane's dilemma.

Suppose selection acts on variation at some locus, so that genotypes *AA*, *Aa*, and *aa* are not all of equal fitness (at least one must be less fit than at least one of the others). The difference will often be minor; perhaps *aa* is only 99% as fit as *AA* or *Aa*. This means that a certain proportion of the population, those with genotype *aa*, are 1% less likely to survive to maturity, or have 1% lower fertility, or in some other way are 1% less effective in leaving offspring. The same or some other pattern of selection may be acting at genetically independent locus *b*, and a small fraction of the population will have reduced fitness from having a less than ideal genotype at the *b*-locus. The argument can be iterated over a possibly enormous number of loci contributing to variation in fitness. This view of selection troubled Haldane because its logical implications seemed incompatible with what normally appears to be happening in real populations.

Most genetic variation is continuous and polygenic. Successful selection for increased yield in a crop plant, for instance, usually causes gene-frequency changes at a number of loci. It is also commonly observed that selection, both artificial and natural, can act on many characters simultaneously. Human evolution over the last few hundred thousand years markedly altered a long list of characters, some of which no doubt required gene-frequency changes at several loci. Haldane assumed that there must be a limit to what selection can accomplish. It can not simultaneously cause finite rates of improvement, or even prevent degeneration, in an infinite number of characters.

Haldane reasoned that if an individual has a less than ideal genotype at *n* loci, it is only $(1-s_1)(1-s_2)\ldots(1-s_n)$ as genetically fit as it ideally might have been. Even an individual with the ideal genotype will be subject to environmental effects on its phenotypic fitness, and even the fittest phenotype is vulnerable to unpredictable hazards. So there must be a finite maximum level of expected success, no matter what the genetic endowment. An individual with a less than ideal genotype at *n* loci would have a mean level of success below that of this maximum. With selection acting on a large number of polymorphic loci, *n* must be large, fitness highly variable, and average fitness extremely low. This is contrary to many biologists' (e.g. Haldane's) impressions of natural populations, although studies of real populations continue to support the prevalence of fitness variation at many loci (Crow and Deniston 1981; Gonzales and Mensua 1987; Innes 1989; M. Lynch 1990).

A bit of arithmetic shows the severity of Haldane's dilemma. There are perhaps 100,000 loci in a mammalian genome (Kauffman 1987), and we might expect at least 10,000 of them to be polymorphic (Ayala 1982), either because one allele is replacing another or because an equilibrium is maintained by mutation pressure, heterosis, or frequency-dependent selection. Perhaps an average individual has a less than ideal genotype at 10% of these loci, or 1000 in all. In a comparable example, Wallace (1989) gives an average individual a suboptimal genotype at 5000 loci. If the average fitness loss per locus is 1% at 1000 loci, the average individual will only be 0.99^{1000}, or about 10^{-5} of the fitness of the best possible genotype. The variance in fitness would be such that almost no individual would be of even 1% of the maximum. Since maximum fitness is no assurance of survival, the population envisioned would have an utterly inadequate gene pool for survival beyond the current generation. The conclusions would of course be more extreme if more than 1% fitness differences were allowed. One-locus fitness differences in nature are often much more than 1% (Endler 1986, pp. 207–11).

Many solutions to the problem were offered during the 1960s and reviewed by Wallace (1970). It was pointed out that different components of fitness need not interact multiplicatively, as my illustration assumes. Having only 90% of maximum resistance to malaria and 90% of maximum resistance to overheating need not mean having 81% as good a chance of surviving both as someone with maximum fitness. If different components of fitness interact synergistically, the selective deaths that take place will be removing a larger proportion of low-fitness genes than they would if the effects on fitness were functionally independent.

This is true, but most of the components of human fitness that we might list, malaria resistance, gamete fertility, sexual attractiveness, visual acuity, fleetness, etc., must vary largely independently, so that multiplicative interaction would be a realistic approximation. There can be only a minor correlation between malaria resistance and visual acuity, for instance, and any such measure of adaptive performance could be a compound of other characters with largely independent effects on fitness. Might not attractiveness to a mosquito interact multiplicatively with immunological and other features to produce whatever resistance to malaria is attained? Even a conceptually simple character, such as grooming behavior, may be a complex of many components, each subject to polygenic inheritance (Vadasz *et al.* 1983). One might plausibly turn the synergism argument around and argue that components of fitness are often compensatory tradeoffs. If famine resistance is correlated with obesity (Knowler *et al.* 1983) it is likely to have a negative correlation with fleetness. Such trade-offs would make genetic load and the cost of natural selection greater than they would be with functional independence.

Wallace (1987, 1989; Reeve *et al.* 1988) recently renewed his effort to lay the problem to rest with arguments and evidence that the traditional genetic-load argument, such as mine above, is based on faulty logic and a misunderstanding of the dynamics of viability selection during the culling of an age cohort. He emphasizes that much culling must take place in every population because of the universal Malthusian factor of over-production of offspring. He envisions a world in which an individual dying as a result of some genetic deficiency is thereby making room for a better endowed individual. If one does not die the other one would. Given that populations do remain finite, the argument fits the facts. Yet the conceptual problem remains, especially in low-fecundity species like our own, because the genetic-load arithmetic makes us expect far more culling than is actually found, and far more than the population could bear.

Wallace's (1987) illustrative model of cohort culling is the competition between seedlings in an experimental tray (e.g. Schmidt and Ehrhardt 1990). The space available will permit only a limited plant biomass to develop, and this will be produced by the small number of successful contenders. The great majority will do very little growing and gradually die out. The possibly small fraction that survives to maturity will be enormously variable in size and fitness. This result seems to be unaffected by levels of genetic load in the seeds used. If a thousand viable seeds are sown in one tray and a thousand with 90% lethal genotypes in another, the two trays may produce about the same number of ultimate survivors, total biomass, and phenotypic fitness variation.

I would suggest another kind of experiment as more relevant to the problem Haldane had in mind. Sow only 100 of the viable seeds in the first tray and 100 with a high incidence of genetic load in the other, and also in each tray sow 900 seeds of competing species. I would expect the 100 viables to win much more representation in their tray than the ten viables and 90 lethals in the other, and this result would be a closer parallel to what usually takes place in the culling of a plant cohort in nature. For most animals, Wallace's experimental model is even less realistic. Only sessile invertebrates meet intense and inescapable competition from near neighbors. Wallace's seedling experiment would be broadly applicable to animal populations only with competition for social status, West Eberhard's (1983) *social selection*, of which sexual selection would be a special case. Social status is a resource that can seldom be appropriated by a member of a different species.

The central problem with Wallace's model, which he calls *soft selection*, is that it implies unrealistically strong density dependence within an age cohort. Most populations in nature are extremely sparse. The tendency for field ecologists to study organisms that are abundant enough to study may greatly bias our impressions. Populations in nature are seldom dense enough to cause any obvious resource depression (Tilman 1982). The most convincing examples of resource depression result from exploitation by many species, such as the reduction of marine invertebrate biomass on mudflats from concerted onslaughts of many species of migratory bird (Schneider 1978). Even the extraordinarily dense populations commonly studied by field ecologists, e.g. by Andrewartha and Birch (1954), usually show numerical changes that look like random fluctuation and seldom give clear evidence of density effects in short-term studies. Individual survival must be mainly a matter of chance, partly a matter of many kinds of adaptive performance, and only to a minor degree affected by density-dependent competition with conspecifics.

Near neighbors, sessile or motile, will often be of different species in diverse natural communities, and the death of one individual will often allow the survival of a member of a competing species. In such situations we most clearly confront the challenge of genetic load in relation to population survival, the problem that worried Haldane. If too large a dose of its population's genetic load causes one individual to die, it is likely to mean that the abundance of that species will be reduced by one. A reduced genetic load would make it more likely for the population to survive in competition with other species. Every $(1-s)$ that enters into a fitness calculation means a finite deficiency in some sort of adaptive performance. Any such deficiency implies not only adverse selection within a population, but also a decreased representation of the population in the community. Dudash's (1990) experiments nicely confirm this expectation. Fitness differences between her inbred and outbred seedlings were much greater in the field than in greenhouse monoculture. The expectation is that the natural populations that are still available for study should be those that have extremely low levels of genetic load, and this is not what is found.

Wallace (1989) claims that Haldane and others have been needlessly worried about a mere 'computational artifact.' They arbitrarily assign a fitness of 1 to a favored genotype and of $1-s$ to an unfavored competitor. If instead we used $1+s$ and 1 we would not calculate such low fitness values for so many multi-locus genotypes. This is true, but the change is merely cosmetic. We would still get the same variation in fitness and be faced with the same problem of how fit the average individual can possibly be. Also, the traditional notation is more realistic. A rare favorable mutation may be said to have a fitness of $1+s$, but this implies a deficiency in the ancestral gene pool. If a mutation can improve some character by some fraction s, that character must have been suboptimal. How could the ancestral population have survived with suboptimal genotypes at a large number of loci if there were competing populations with a lower genetic load?

I think the time has come for renewed discussion and experimental attack on Haldane's dilemma.

10.2 Paradoxes of sexuality

Sexual reproduction by its existence and in many special aspects is a complex of puzzles on which many books have been written (e.g. Bradbury and Andersson 1987; Stearns 1987; Michod and Levin 1988). The main theoretical challenge is in the *cost of meiosis*, but this is a matter already getting attention from able investigators. I can do no better than refer readers to Maynard Smith (1984*b*), Eberhard (1985), Felsenstein (1985*a*), Bierzychudek (1989), Hamilton *et al.* (1990), Parts II and III of Stearns (1987) and Chapters 4–9 in Michod and Levin (1988). Another major challenge is in resolving the data of life-history diversity in the frequency and the developmental and ecological correlates of sexual phases. The problem here is not so much logical as logistic. The diversity is overwhelming in relation to the time and money that a few thousand interested biologists can devote to it.

Of the many recombination-related difficulties that I could discuss I will echo Maynard Smith (1988*a*) and choose the one that best serves as a kind of text-book illustration of an evolutionary anomaly, the absence of sexual reproduction throughout the rotifer order Bdelloidea. This is anomalous because it clearly violates the principle of Muller's ratchet, which seems a logically tight line of reasoning from well established premises (but see Gabriel (1987)). Muller (1964) was the first to recognize that an asexual lineage 'incorporates a kind of ratchet mechanism, such that it can never get to contain, in any of its lines, a load of mutation smaller than that already existing in its at present least loaded lines.' It can acquire a higher load of mutation simply by the occurrence of a new one in a least loaded line. So exclusively asexual reproduction leads inevitably to a degeneration of the genome, in the sense of its being ever more ruled by chemical stability, and ever less informative as to what has succeeded in the past. This must always lead to rapid extinction on an evolutionary time scale. For a recent quantitative study of Muller's ratchet, see G. Bell (1988).

Muller's ratchet explains the phylogenetic distribution of asexual species in most major groups of eukaryotes. There is a fair number of exclusively clonal species, but never any entirely clonal genera or higher categories. Asexual species arise from time to time, but Muller's ratchet must lead them to extinction long before they can produce any appreciable taxonomic diversification. The loss of sexuality seems to be a classic example of an evolutionary step that is opposed by clade selection (Van Valen 1975). Unfortunately the general rule of conformity to expectations of Muller's ratchet has some exceptions. The whole rotifer order Bdelloidea (Meglitch (1967) calls them a class), with its several families and many genera and species, is composed entirely of parthenogenetic females (Pennak 1978). There are also a few other noteworthy violators of the theory, such as the freshwater gastrotrich order Chaetonotoidea (Meglitch 1967; Pennak 1978).

Another difficulty that surely deserves more attention is the scarcity of adaptively flexible sex determination (for a comprehensive review, see Bull, (1983)). Sex determination in most animals is genetic and is fixed at conception. Only a few have sex determination as a facultative response to information perceived during development. A neatly understandable example (Conover and Heins 1987) is provided by a fish, the Atlantic silverside, which spawns every spring in shallow waters along the Atlantic coast of the United States and Canada. Eggs spawned early in the season become females; those later on become males. Temperature provides the cue, so that development below a certain threshold causes female development, warmer water male development. This mechanism gives females a longer growing season and larger size for the following spring's spawning. The close relationship between size and fecundity in fishes makes large size more important to female fitness than to male.

This environmental sex determination is adaptive in a way that extends the size-advantage model used to explain the occurrence of protandry vs. protogyny among sequential hermaphrodites (Ghiselin 1969; Warner 1975). The silverside is almost entirely semelparous, and this would rule out sequential hermaphroditism as a viable life history. As predicted (Conover and Heins 1987), the temperature threshold that determines sex varies as expected of combined optimizing and frequency-dependent selection. It is lower in northern, higher in southern parts of the range, so that the sexes are nearly equally abundant all along the coast. Experiments by Conover and Van Voorhees (1990) show that the threshold can be changed by selection in the laboratory.

Besides the greater dependence of reproductive success on size in females than in males in most animals, it is possible to think of many other ways in which it may be more adaptive for an individual in a given situation to be male or female, and to identify cues that would predict the situation during development. A clear example would be the stochastically varying sex ratios of many social groups. In a pond in which most of the frogs happen to be of sex A, it would pay a tadpole to develop into a member of sex B. It would also pay a parent to bias sex determination away from whatever is the majority in previously produced young (Taylor and Sauer 1980). This would avoid what might be called *Baptista's Burden*. Having Bianca instead of a son after having had Katharina was something of a challenge to his fitness. Yet despite the advantage that can be envisioned in alternating sons and daughters, each sex determination is a largely independent event in most animal populations (Williams 1979; Huck *et al.* 1990).

Facultative sex determination would only be expected in groups in which useful cues can be perceived prior to any major developmental commitment to maleness or femaleness, requirements discussed in detail by Bull (1983), Charnov and Bull (1977), and Korpelainen (1990). These conditions must surely obtain in many diploid insect populations. Why is adaptive sex determination, such as that found in the silverside, not widespread in many groups of insects? Similar arguments can be made for parental manipulation of offspring sex ratio, as adjustments either to the conditions to be faced by the young, or to varying parental aptitude for raising sons or daughters. Much theoretical work has been done since the pioneering efforts by Kolman (1960) and Verner (1965), but the expected adaptations are infrequent and weakly developed (Clutton-Brock and Iason 1986; Huck *et al.* 1990), except in haplodiploid animals (Charnov 1982, pp. 48–66). Even in groups in which environmental cues determine sex, for instance the reptiles, the responses may make little adaptive sense to the leading investigators of these phenomena (Bull and Charnov 1989; Charnov and Dawson 1989).

10.3 Other difficulties

The problem of the evolution of senescence is solved, in a general and qualitative way (Rose 1991), and no recent work has improved on that of Hamilton (1966). Improvement is long overdue. We need an explicit theorem of the form

$$y_x = f(\hat{y}, x, u, v, w \ldots)$$

where y_x is a measure of adaptive performance at age x, like fleetness, visual acuity, or the ability to avoid death. The constant $\hat{y}$ would be the maximum value of y. The independent variables would include age (x) and perhaps age-specific survivals, fertilities, and other factors. The equation needs to be derived logically from a fitness-maximization model with realistic genetic, developmental, and demographic constraints. It must then be shown to predict successfully the effects of age on measures of adaptive performance in a diversity of populations subject to senescence.

It will be necessary to allow the derivation to include at least some numerical methods and algebraic approximations, because some of the likely causal factors, such as survivorship, depend recursively on all previous levels of adaptive performance. This rules out any purely analytic solution to the problem. One straightforward approach would assume that ability to avoid death at a given age is proportional to the strength of selection for fitness at that age. This would be determined by the product of survivorship and reproductive value. If so it should be possible to derive, by simulation, an age-distribution of mortality rates and from them a theoretical age structure that can be compared to observed age structures. This approach fails utterly (Williams and Taylor 1987). It predicts far more abrupt senescence than is actually found, for instance that human survival beyond 60 years is impossible, even with unrealistically liberal estimates of kin-selected contributions from what have normally been classified as post-reproductive age groups.

A related problem is the unexplained but now undeniable difference in age structures of avian and mammalian populations (Nesse 1988). Both birds and mammals have life cycles that should make them similarly vulnerable to the evolution of senescence, but there is little evidence that senescence affects birds at all. Where data on avian age structures are most abundant, it usually appears that mortality rates of young adults prevail throughout adult life. This conspicuously violates expectation from theory (of Hamilton 1966).

I offer the following weak rationalization of this contradiction. Birds really do senesce, with the same general pattern as that found in mammals, but much more slowly. A 50-year-old robin really would be afflicted with old-age impairments, but no such bird will ever be found in nature. Mortality rates of young adults are high enough to make it extremely unlikely that any individual would live to even half that age. If so the problem is merely to explain why birds should deteriorate so much more slowly than mammals with similar life histories.

The only major developmental difference in the life cycles of birds and mammals is in the relation between age of sexual maturation and age of attainment of full size and somatic capability (Fig. 10.1). Mammals typically become sexually mature a bit before full size is reached. Birds may reach full size in a few weeks but then delay reproduction until they are at least one, and perhaps several years of age. Senescence theoretically begins at sexual maturity. This means that a mammal's somatic machinery can start deteriorating even before it is fully formed. A bird's somatic machinery is selected to perform with maximal effectiveness for many weeks, perhaps years after it is full grown.

An analogy to the mammal would be a tool sold with a strange sort of warranty. It must perform perfectly in the shop before purchase, and thereafter a purchaser is entitled to a fractional refund for the tool's failure, the fraction starting high and slowly approaching zero as time goes on. An analogy to the bird would be the same sort of tool with a

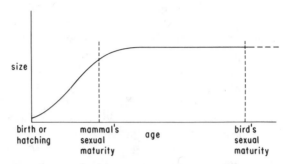

Fig. 10.1. Growth and maturity of typical birds and mammals. I suggest that delaying sexual maturity until long after the somatic machinery is fully formed has the incidental effect of retarding a bird's senescence relative to a mammal's.

year's full warranty. Only after a year would the fractional-refund rule go into effect. My guess is that if these tools are well designed from the seller's perspective, the bird-tool will not only be more likely than the mammal-tool to last at least a year, but also likely to have a lower rate of deterioration thereafter.

Many other problems remain. Discussions of the evolution of the human brain and behavior have lain largely in the purview of philosophers, or of anthropologists and biologists in a philosophical mood. Special aspects of it have recently had serious attention from a biological perspective: the possibility of innate problem-solving algorithms (Cosmides 1985), of language-acquisition machinery (Chomsky 1986; Pinker and Bloom 1989), of biological factors in sexual and parental motivation (Symons 1979; Hrdy 1981), of the evolution of morality (Humphrey 1976; Alexander 1987). The human brain is not just a unique challenge to biologists, but a matter of universal intellectual interest deserving of well-disciplined thought and well-supported research. The insights produced can be expected to have many medical, economic, educational, and other practical applications.

Increased concern about organisms or special adaptations that we might think ought to exist, but do not, such as adaptation of mammalian spermatogenesis to slightly higher temperature (Chapter 9), can be expected to suggest new ideas about internal constraints and external evolutionary pressures. A human eye blink takes about 50 milliseconds. This means that we are blind about 5% of the time when we are using our eyes normally. Many events of importance can happen in 50 milliseconds, so that we might often miss them entirely. A rock or spear thrown by a powerful adversary can travel more than a meter in 50 milliseconds, and it could be important to perceive such motion as accurately as possible. Why then do we blink with both eyes simultaneously? Why not alternate and replace 95% visual attentiveness with 100%? I can imagine an answer in some sort of trade-off balance. A blink mechanism for both eyes at once may be much simpler and cheaper than one that regularly alternates. The visual gain from regular alternation might not be worth whatever additional costs might be required. I do not see how synchrony could have any such advantage over an independent blinking that would soon drift out of phase.

Am I assigning too much importance to 50 milliseconds of blindness every second? I think not, after reading Muller's (1948) 'Evidence of the precision of genetic adaptation' and accounts of the intricate mechanisms of the visual system for extracting every possible usefulness from optical stimuli (Ali and Klyne 1985). Serious consideration of why natural selection permits simultaneous blinking might yield otherwise elusive insights. What change in the machinery would be needed to produce the first step towards my envisioned adaptive alternation or simple independent timing? How might the change be achieved developmentally? What other changes would be expected from a mutation that produced a slight lag in the blinking of one eye? How would selection act on such a mutation?

I have no reason to believe that Vermeij (1987, p. 26) would approve my belaboring his wish to know why certain kinds of organisms, perhaps mammals with asynchronous blinking, do not exist. It does appear that his wish and perhaps my elaboration would have pleased R. A. Fisher (1930, pp. viii–ix) who quoted with approval a mathematician's statement

We need scarcely add that the contemplation in natural science of a wider domain than the actual leads to a far better understanding of the actual.

and then lamented the fact that, in his time,

No practical biologist interested in sexual reproduction would be led to work out the detailed consequences experienced by organisms having three or more sexes: yet what else should he do if he wishes to understand why the sexes are, in fact, always two?

I am sure that the next few years will provide many fine studies, theoretical and empirical, of adaptations shown by organisms. I hope they also show increased attention to the adaptations not shown by organisms.

References

Alexander, R. D. (1981). Evolution, culture, and human behavior: Some general considerations. In *Natural selection and social behavior* (ed. R. D. Alexander and D. W. Tinkle), pp. 509–20, Chiron Press, New York.

Alexander, R. D. (1987). *The biology of moral systems*. Aldine de Gruyter, New York.

Allee, W. C., Emerson, A. E., Park, T., Park, O. and Schmidt, K. P. (1948). *Principles of animal ecology*. Saunders, Philadelphia.

Ali, M. A. and Klyne, M. A. (1985). *Vision in vertebrates*. New York, Plenum.

Amadjian, V. (1990). What have synthetic lichens told us about real lichens? *Bibliography of Lichenology* **38**, 3–12.

Anderson, D. J. (1990). On the evolution of human brood size. *Evolution* **44**, 438–40.

Anderson, D. J., Stoyan, N.C. and Ricklefs, R.E. (1987). Why are there no viviparous birds? A comment. *American Naturalist* **130**, 941–7.

Andersson, M. (1982). Female choice selects for extreme tail length in a widowbird. *Nature* **299**, 818–20.

Andren, C., Marden, M. and Nilson, G. (1989). Tolerance to low pH in a population of moor frogs, *Rana arvalis* from an acid and a neutral environment a possible test case of rapid evolutionary response to acidification. *Oikos* **56**, 215–23.

Andrewartha, H. G. and Birch, L. C. (1954). *The distribution and abundance of animals*. University of Chicago Press.

Arthur, W. (1988). *A theory of the evolution of development*. Wiley, New York.

Ashmole, N. P. (1971) Sea bird ecology and the marine environment. In *Avian biology* (ed. D. S. Farner, J. R. King and K. C. Parkes), Vol. 1, pp. 223–86, Academic Press, New York.

Avers, C. J. (1989). *Process and pattern in evolution*. Oxford University Press, New York.

Ayala, F. J. (1968). Biology as an autonomous science. *American Scientist* **56**, 207–21.

Ayala, F. J. (1982). The genetic structure of species. In *Perspectives on evolution* (ed. R. Milkman), pp.60–82, Sinauer, Sunderland, MA.

Barash, D. P. (1982). *Sociobiology and behavior*. 2nd edition. Elsevier, New York.

Barber, N. (1991). Play and energy regulation in mammals. *Quarterly Review of Biology* **66**, 129–47.

Barkai, A. and McQuaid, C. (1988). Predator–prey role reversal in a marine benthic ecosystem. *Science* **242**, 62–4.

Barkan, C. P. L. (1990). A field test of risk-sensitive foraging in black-capped chickadees (*Paris atricapillus*). *Ecology* **71**, 391–400.

Barnes, B. M. (1989). Freeze avoidance in a mammal: Body temperature below 0°C in an Arctic hibernator. *Science* **244**, 1593–5.

Barrett, S. C. H. (1989). Mating system evolution and speciation in heterostylus plants. In *Speciation and its consequences* (ed. D. Otte and J. A. Endler), pp.257–83, Sinauer, Sunderland, MA.

Barrett, S. C. H., Morgan, M. T. and Husband, B. C. (1989). The dissolution of a complex genetic polymorphism: The evolution of self-fertilization in tristylous *Eichhornia paniculata* (Pontederiaceae). *Evolution* **43**, 1398–1416.

Barton, N. H. (1989). Founder effect speciation. In *Speciation and its consequences* (ed. D. Otte and J. A. Endler), pp. 229–56, Sinauer, Sunderland, MA.

Barton, N. H. and Hewett, G. M. (1985). Analysis of hybrid zones. *Annual Review of Ecology and Systematics* **16**, 113–148.

Bates, D. E. B. and Kirk, N. H. (1985). Graptolites, a fossil case-history of evolution from sessile, colonial animals to automobile superindividuals. *Proceedings of the Royal Society of London, B* **228**, 207–24.

Baumgartner, J. V., Bell, M. A. and Weinberg, P. H. (1988). Body form differences between the Enos Lake species pair of threespine sticklebacks (*Gasterosteus aculeatus* complex). *Canadian Journal of Zoology*, **65**, 467–74.

Beamish, R. J. (1987). Evidence that parasitic and nonparasitic life history types are produced by one population of lamprey. *Canadian Journal of Fisheries and Aquatic Sciences* **44**, 1779–82.

Bell, G. (1988). Recombination and the immortality of the germ line. *Journal of Evolutionary Biology* **1**, 67–72.

Bell, G. (1989). A comparative method. *American Naturalist* **133**, 553–71.

Bell, M.A. (1989). Stickleback fishes: Bridging the gap between population biology and paleobiology. *Trends in Ecology and Evolution* **3**, 320–5.

Belovsky, G. E. (1984). Summer diet optimization by beaver. *American Midland Naturalist* **111**, 209–22.

Belsky, A. J. (1986). Does herbivory benefit plants? A review of the evidence. *American Naturalist* **127**, 870–92.

Belsky, A. J. (1987). The effects of grazing: Confounding of ecosystem, community, and organism scales. *American Naturalist* **129**, 777–83.

Benton, M. J. (1987). Progress and competition in macroevolution. *Biological Review* **62**, 305–38.

Bentzen, P., Ridgway, M. S. and McPhail, J. D. (1984). Ecology and evolution of sympatric sticklebacks (*Gasterosteus*): spatial segregation and seasonal habitat shifts in the Enos Lake species pair. *Canadian Journal of Zoology* **62**, 2436–9.

Bernstein, J. W. and Smith, R. J. F. (1983). Alarm substance cells in fathead minnows do not affect the feeding preference of rainbow trout. *Environmental Biology of Fishes* **9**, 307–11.

Best, R. C. (1981). The tusk of the narwhal (*Monodon monoceros* L.): interpretation of its function (Mammalia: Cetacea). *Canadian Journal of Zoology* **59**, 2386–93.

Betzig, L. L., Borgerhoff Mulder, M. and Turke, P. W. (1987). *Human reproductive behavior*. Cambridge University Press.

Bierzychudek, P. (1989). Environmental sensitivity of sexual and apomictic *Antennaria*: Do apomicts have general-purpose genotypes? *Evolution* **43**, 1456–66.

Billington, H.L., Mortimer, A.M. and McNeilly, T. (1988). Divergence and genetic structure in adjacent grass populations. *Evolution* **42**, 1267–77.

Birch, M. (1974). Aphrodisiac pheromones in insects. In *Pheromones* (ed. M. C. Birch), pp. 115–34, North-Holland, Amsterdam.

Blackburn, D. G. and Evans, H. E. (1986). Why are there no viviparous birds? *American Naturalist* **128**, 165–90.

Blaustein, A. R., Porter, R. H. and Breed, M. D. (editors) (1988). Special issue: Kin recognition in animals. *Behavior Genetics* **18**, 405–82.

Boake, C. R. B. (1986). A method for testing adaptive hypotheses of mate choice. *American Naturalist* **127**, 654–66.

Bock, W. J. (1986). Species concepts, speciation, and macroevolution. In *Modern aspects of species* (ed. K. Iwatsuki, P. Raven and W. J.Bock), pp. 31–57, University of Tokyo Press.

Boetius, J. and Harding, E. F. (1985). A re-examination of Johannes Schmidt's Atlantic eel investigations. *Dana* **4**, 129–62.

Bond, W.J. (1989). The tortoise and the hare: ecology of angiosperm dominance and gymnosperm persistence. *Biological Journal of the Linnaean Society* **36**, 227–49.

Bonhomme, F. (1986). Molecules, populations and species evolution in the genus *Mus* (Mammalia: Rodentia). In *Modern aspects of species* (ed. K. Iwatsuki, P.Raven and W. J. Bock), pp. 125–43, University of Tokyo Press.

Bonner, J. T. (1980). *The evolution of culture in animals*. Princeton University Press.

Bonner, J. T. (1988). *The evolution of complexity*. Princeton University Press.

Booth, E. S. and Chiasson, R. B. (1967). *Laboratory anatomy of the cat*. 4th edition, W. C. Brown, Dubuque.

Borgia, G. (1979). Sexual selection and the evolution of mating systems. In *Sexual selection and reproductive competition in insects* (ed. M. S. Blum and N. A. Blum), pp. 19–80, Academic Press, New York.

Boucher, D. H. (editor) (1985). *The biology of mutualism.* Oxford University Press, New York.

Boyce, M. S. and Perrins, C. M. (1987). Optimizing great tit clutch size in a fluctuating environment. *Ecology* **68**, 142–53.

Boyd, R. and Richerson, P. J. (1985). *Culture and the evolutionary process.* University of Chicago Press.

Bradbury, J. W. and Andersson, M. B. (1987). *Sexual selection: Testing the alternatives.* (Dahlem: Life Sciences Research Report, 39), Wiley, Chichester.

Bradbury, J. W. and Gibson, R. M. (1983). Leks and mate choice. In *Mate choice* (ed. P. Bateson), pp. 109–38, Cambridge University Press.

Bradbury, J. W., Vehrencamp, S. L. and Gibson, R. (1985). Leks and the unanimity of female choice. In *Evolution. Essays in honor of John Maynard Smith.* (ed. P. H. Greenwood and P. H. Harvey), pp. 301–14, Cambridge University Press.

Brande, S. (1979). Biometric analysis and evolution of two species of *Mulinia* (Bivalvia: Mactridae) from the late Cenozoic of the Atlantic coastal plain. Ph.D dissertation. State University of New York, Stony Brook.

Brandon, R. N. (1988). Levels of selection: A hierarchy of interactors. In *The role of behavior in evolution*, (ed. H.C. Plotkin), pp. 51–71, MIT Press, Cambridge, MA.

Brandon, R. N. (1990). *Adaptation and environment.* Princeton University Press.

Breder, C. M., Jr. and Rosen, D. E. (1966). *Modes of reproduction in fishes.* Natural History Press, Garden City.

Briand, F. and Cohen, J. E. (1984). Community food webs have scale-invariant structure. *Nature* **307**, 264–7.

Brodie, E. D., III and Brodie, E. D., Jr. (1991). Evolutionary responses of predators to dangerous prey: Reduction of toxicity of newts and resistence of garter snakes in island populations. *Evolution* **45**, 221–4.

Brown, J. H. (1987). Variation in desert rodent guilds: Patterns, processes, and scales. In *Organization of communities past and present*, The 27th Symposium of the British Ecological Society (ed. J. H. R. Gee and P. S. Giller), pp.185–203, Blackwell Scientific Publications, Oxford.

Brown, W. L, Jr. (1958). General adaptation and evolution. *Systematic Zoology* **7**, 157–68.

Bull, J. J. (1983). *Evolution of sex determining mechanisms.* Benjamin/Cummings, Menlo Park.

Bull, J. J. and Charnov, E. L. (1985). On irreversible evolution. *Evolution* **39**, 1149–55.

Bull, J. J. and Charnov, E. L. (1989). Enigmatic reptilian sex ratios. *Evolution* **43**, 1561–6.

Bull, J. J., Molineux, I. J. and Rice, W. R. (1991). Selection of benevolence in a host-parasite system. *Evolution* **45**, 875–89.

Bulmer, M. G. (1970) *The biology of twinning in Man*. Clarendon, Oxford.

Bulmer, M. (1988). Sex ratio evolution in lemmings. *Heredity* **61**, 231–3.

Burghardt, G. M. (1970) Defining 'communication.' In *Communication by chemical signals* (ed. J. W. Johnson, D. G. Moulton and A. Turk), pp. 5–18, Appleton, New York.

Burley, N. (1986) Sexual selection for aesthetic traits in species with biparental care. *American Naturalist* **127**, 415–45.

Buss, L. W. (1987). *The evolution of individuality*. Princeton University Press.

Cain, A. J. (1964). The perfection of animals. In *Viewpoints in biology* (ed. J. D. Carthy and C. L. Duddington), Vol. 3, pp. 36–63 , Butterworth, London.

Carde, R. T. and Charleton, R. E. (1984). Olfactory sexual communication in Lepidoptera: Strategy, sensitivity, and selectivity. In *Insect communication* (ed. T. Lewis), pp. 241–65, Academic Press, London.

Carlander, K. D. (1969–77). *Handbook of freshwater fishery biology*. Vol. 1 (1969), Vol. 2 (1977), Iowa State University Press, Ames.

Carroll, R. L. (1988). *Vertebrate paleontology and evolution*. Freeman, New York.

Carson, H. L. (1989). Genetic imbalance, realigned selection, and the origin of species. In *Genetics, speciation, and the Founder principle* (eds. L. V. Giddings, K. Y. Kanishiro and W. W. Anderson), pp. 345–62, Oxford University Press, New York.

Cavalli-Sforza, L. L. and Feldman, M. W. (1981). *Cultural transmission and evolution: A quantitative approach*, Monographs in population biology 16, Princeton University Press.

Charlesworth, B. (1980). *Evolution in age-structured populations*. (Cambridge studies in mathematical biology), Cambridge University Press, New York.

Charlesworth, B. (1987). The heritability of fitness. In *Sexual selection: testing the alternatives* (ed. J.W. Bradbury and M. B. Andersson), pp. 21–40, Wiley, Chichester.

Charlesworth, B. and Rouhani, S. (1988). The probability of peak shifts in a founder population. II. An additive polygenic trait. *Evolution* **42**, 1129–45.

Charnov, E. L. (1976). Optimal foraging: the marginal value theorem. *Theoretical Population Biology* **9**, 129–36.

Charnov, E. L. (1982). *The theory of sex allocation*. Princeton University Press.

Charnov, E. L. (1989). Phenotypic evolution under Fisher's fundamental theorem of natural selection. *Heredity* **62**, 113–16.

Charnov, E. L. and Bull, J. (1977). When is sex environmentally determined. *Nature* **266**, 828–30.

Charnov, E. L. and Dawson, T. D. (1989). Environmental sex determination with overlapping generations. *American Naturalist* **134**, 806–16.

Chomsky, N. (1986). *Knowledge of language: Its nature, origin, and use*. Praeger, New York.

Chow, S., Fujio, Y. and Nomura, T. (1988). Reproductive isolation and distinct population structures in two types of the freshwater shrimp *Palaemon paucidens*. *Evolution* **42**, 804–13.

Churchland, P. S. (1986). *Neurophysiology. Toward a unified science of the mind-brain*. MIT Press, Cambridge, MA.

Clark, A. B. (1991). Individual variation in responsiveness to environmental change. In *Primate responses to environmental change* (ed. H. O. Bon), pp. 91–110. Chapman and Hall, London.

Clark, A. B. and Ehlinger, T. J. (1987). Pattern and adaptation in individual behavioral differences. *Perspectives in Ethology* **7**, 1–47.

Clausen, J., Keck, D. D. and Hiesey, W. M. (1941). Regional differentiation in plant species. *American Naturalist* **75**, 231–50.

Clausen, J., Keck, D. D. and Hiesey, W. M. (1947). Heredity of geographically and ecologically isolated races. *American Naturalist* **81**, 114–33.

Clutton-Brock, T. H. (1988). *Reproductive success: studies of individual variation in contrasting breeding systems*. University Chicago Press.

Clutton-Brock, T. H. and Iason, G. R. (1986). Sex ratio variation in mammals. *Quarterly Review of Biology* **61**, 339–74.

Cockburn, A. (1988). *Social behaviour in fluctuating populations*. Croom Helm, London.

Coe, W. R. (1949). Divergent methods of development in morphologically similar species of prosobranch gastropods. *Journal of Morphology* **84**, 383–99.

Coleman, R. M., Gross, M. R. and Sargent, R. C. (1985). Parental investment decision rules: a test in bluegill sunfish. *Behavioral Ecology and Sociobiology* **18**, 59–66.

Colgan, P. (1989). *Animal motivation*. Chapman and Hall, New York.

Colwell, R. K. (1986). Population structure and sexual selection for host fidelity in the speciation of hummingbird flower mites. In *Evolutionary processes and theory* (ed. S. Karlin and E. Nero), pp. 475–95, Academic Press, Orlando.

Conover, D. O. and Heins, S. W. (1987). The environmental and genetic components of sex ratio in *Menidia menidia* (Pisces: Atherinidae). *Copeia* (3), 732–43.

Conover, D. O. and Van Voorhees, D. A. (1990). Evolution of a balanced sex ratio by frequency-dependent selection in a fish. *Science* **250**, 1556–8.

Cooch, E. G., Lank, D. B., Rockwell, R. F. and Cooke, F. (1989). Long term decline in fecundity in a snow goose population: evidence for density dependence? *Journal of Animal Ecology* **58**, 711–26.

Cook, R. E. (1983). Clonal plant populations. *American Scientist* **71**, 244–53.

Cooke, F. (1988). Genetic studies of birds—the goose with blue genes. *Proceedings of the International Ornithological Congress* **19**, 189–214.

Cooke, F., Findlay, C. S. and Rockwell, R. F. (1984). Recruitment and the timing of reproduction in lesser snow geese (*Chen caerulescens caerulescens*). *Auk* **101**, 451–8.

Cooke, F., Taylor, P. D., Francis, C. M. and Rockwell, R. F. (1990). Directional selection and clutch size in birds. *American Naturalist* **136**, 261–67.

Cooper, W. E., Jr., Gartska, W. R., and Vitt, L. J. (1986). Female sex pheromone in the lizard *Eumeces laticeps*. *Herpetologia* **42**, 361–6.

Corbetta, M., Miezin, F. M., Dobmeyer, S., Shulman, G. L. and Petersen, S. E. (1990). Attentional modulation of neural processing of shape, color, and velocity in humans. *Science* **248**, 1556–9.

Cosmides, L. M. (1985). Deduction or Darwinian algorithms? An explanation of the 'elusive' content effect on the Wason selection task. Ph.D. dissertation. Harvard University.

Cosmides, L. M. and Tooby, J. (1981). Cytoplasmic inheritance and intergenomic conflict. *Journal of Theoretical Biology* **89**, 83–129.

Cracraft, J. (1989). Speciation and its ontology: The empirical consequences of alternative species concepts for understanding patterns and processes of differentiation. In *Speciation and its consequences* (ed. D. Otte and J. A. Endler), pp. 27–59, Sinauer, Sunderland, MA.

Cracraft, J. (1990). The origin of evolutionary novelties: Pattern and process at different hierarchical levels. In *Evolutionary innovations* (ed. M. H. Nitecki), pp. 21–44, University of Chicago Press.

Crocker, G. and Day, T. (1987). An advantage to mate choice in the seaweed fly, *Coelopa frigida*. *Behavioral Ecology and Sociobiology* **20**, 295–301.

Crow, J. F. (1987). Neutral models in molecular evolution. In *Neutral models in biology* (ed. M. H. Nitecki and A. Hoffman), pp. 11–22, Oxford University Press, New York.

Crow, J. F. (1988). Progress in genetics [book review]. *Science* **242**, 1449–50.

Crow, J. F. and Denniston, C. (1981). The mutation component of genetic damage. *Science* **212**, 888–93.

Crowl, T. A. and Covich, A. P. (1990). Predator-induced life-history shifts in a freshwater snail. *Science* **247**, 949–51.

Cuthill, I. C., Kacelnik, A., Krebs, J. R., Haccou, P. and Iwasa, Y. (1990). Starlings exploiting patches: the effect of recent experience on foraging behavior. *Animal Behavior* **40**, 625–40.

Damuth, J. (1985). Selection among 'species': A formulation in terms of natural functional units. *Evolution* **39**, 1132–46.

Damuth, J. and Heisler, I. L. (1988). Alternative formulations of multilevel selection. *Biology and Philosophy* **3**, 407–30.

Darwin, C. R. (1859). *The origin of species*. John Murray, London.

Davies, N. B. and Brooke, M. de L. (1989). An experimental study of co-evolution between the cuckoo, *Cuculus canorus*, and its hosts. II. Host egg markings, chick discrimination and general discussion. *Journal of Animal Ecology* **58**, 225–36.

Davis, M. B. (1989). Insights from paleoecology on global change. *Bulletin of the Ecological Society of America* **70**, 222–8.

Davis, W. P. (1988). Reproductive and developmental responses of the self-fertilizing fish, *Rivulus marmoratus*, induced by the plasticizer, di-n-butylphthalate. *Environmental Biology of Fishes* **21**, 81–90.

Davison, G. W. H. (1983). The eyes have it: ocelli in a rainforest pheasant. *Animal Behavior* **31**, 1037–42.

Dawkins, R. (1976). *The selfish gene*. Oxford University Press, New York.

Dawkins, R. (1982*a*). *The extended phenotype*. W.H. Freeman, Oxford.

Dawkins, R. (1982*b*). Replicators and vehicles. In *Current problems in sociobiology* (ed. King's College Sociobiology Group), pp. 45–64, Cambridge University Press.

Dawkins, R. (1983). Universal Darwinism. In *Evolution from molecules to men*. (ed. D.S. Bendall), pp. 203–25, Cambridge University Press.

Dawkins, R. (1986). Wealth, polygyny, and reproduction success. *Behavior and Brain Science* **9**, 190–1.

Dawkins, R. (1989). Why any study of human origins must be Darwinian. In *Human origins* (ed. J. R. Durant), pp. 1–8, Oxford University Press.

de Jong, G. (1987). Consequences of a model of counter-gradient selection. In *Population genetics and evolution* (ed. G. de Jong), pp. 264–77, Springer-Verlag, Berlin.

Demerec, M. (1951) Studies of the streptomycin-resistance system of mutations in *E. coli. Genetics* **36**, 585–97.

Denning, P. J. (1990) Is thinking computable? *American Scientist* **78**, 100–2.

Dial, K. P. and Marzluff, J. M. (1989). Nonrandom diversification within taxonomic assemblages. *Systematic Zoology* **38**, 26–37.

Dobzhansky, T. (1941). *Genetics and the origin of species*, 2nd edition. Columbia University Press, New York.

Dodds, W. K. (1988). Community structure and selection for positive or negative species interactions. *Oikos* **53**, 387–90.

Dominey, W. J. (1981). Maintenance of female mimicry as a reproductive strategy in the bluegill sunfish (*Lepomis macrochirus*). *Environmental Biology of Fishes* **6**, 59–64.

Dominey, W. J. (1983). Mobbing in colonial nesting fishes, especially in the bluegill, *Lepomis macrochirus*. *Copeia* (4), 1086–8.

Doolittle, W. F. and Sapienza, C. (1980). Selfish genes, the phenotype paradigm and genome evolution. *Nature* **284**, 601–3.

Doust, J. L. (1980). Experimental manipulations of patterns of resource allocation in the growth cycle of *Smyrnium olusatrum* L. *Biological Journal of the Linnaean Society* **13**, 155–66.

Dowdey, T. G. and Brodie, E. D., Jr. (1989). Antipredator strategies of salamanders: individual and geographical variation in responses of *Eurycea bislineata* to snakes. *Animal Behavior* **38**, 707–11.

Drake, J. A. (1991). Community-assembly mechanisms and the structure of an experimental species ensemble. *American Naturalist* **137**, 1–26.

Dretske, F. I. (1981). *Knowledge and the flow of information*. Basil Blackwell, Oxford.

Dretske, F. I. (1985). Precis of *Knowledge and the Flow of Information*. In *Naturalizing epistemology* (ed. H. Kornblith), pp. 169–87, MIT Press, Cambridge, MA.

Driesch, H. (1929). *The science and philosophy of the organism*, 2nd edition. A & C Black, London.

Dudash, M. R. (1990). Relative fitness of selfed and outcrossed progeny in a self-compatible, protandrous species, *Sabatia angularis* L. (Gentianaceae): A comparison in three environments. *Evolution* **44**, 1129–39.

Dunbrack, R. L. and Ramsay, M. A. (1989). The evolution of viviparity in amniote vertebrates: Egg retention versus egg size reduction. *American Naturalist* **133**, 138–48.

Eberhard, W. G. (1980). Evolutionary consequences of intracellular organelle competition. *Quarterly Review of Biology* **55**, 231–49.

Eberhard, W. G. (1985). *Sexual selection and animal genitalia*. Harvard University Press, Cambridge, MA.

Eberhard, W. G. (1990a). Evolution in bacterial plasmids and levels of selection. *Quarterly Review of Biology* **65**, 3–22.

Eberhard, W. G. (1990b). Animal genetalia and female choice. *American Scientist* **78**, 134–41.

Echelle, A. A. and Connor, P. J. (1989). Rapid, geographically extensive genetic introgression after secondary contact between two pupfish populations (Cyprinodon, Cyprinodontidae). *Evolution* **43**, 717–27.

Ehlinger, T. J. (1989). Learning and individual variation in bluegill foraging: habitat specific techniques. *Animal Behavior* **38**, 643–58.

Ehlinger, T. J. (1990). Habitat choice and phenotype-limited feeding efficiency in bluegill: Individual differences and trophic polymorphism. *Ecology* **71**, 886–96.

Ehlinger, T. J. and Wilson, D. S. (1988). Complex foraging polymorphism in bluegill sunfish. *Proceedings of the National Academy of Sciences* **85**, 1878–82.

Eldredge, N. (1989). *Macroevolutionary dynamics: Species, niches, and adaptive peaks*. McGraw-Hill, New York.

Eldredge, N. and Gould, S. J. (1972). Punctuated equilibria: an alternative to phyletic gradualism. In *Models in Paleobiology* (ed. T.J.M. Schopf), pp. 82–115, Freeman Cooper, San Francisco.

Eldredge, N. and Gould, S. J. (1988). Punctuated equilibrium prevails. *Nature* **332**, 211–12.

Endler, J. A. (1977). *Geographic variation, speciation, and clines*. Princeton University Press.

Endler, J. A. (1986). *Natural selection in the wild*. Princeton University Press.

Eschelle, A. A. and Connor, P. J. (1989). Rapid, geographically extensive genetic introgression after secondary contact between two pupfish species (*Cyprinodon*, Cyprinodontidae). *Evolution* **43**, 717–27.

Evans, D. A., Baker, R. and Howse, P. E. (1979). The chemical ecology of termite defense behaviour. In *Chemical ecology: odour communication in animals* (ed. F. J. Ritter), pp. 213–24, Amsterdam, Elsevier/North Holland.

Ewald, P. W. (1988). Cultural vectors, virulence, and the emergence of evolutionary epidemiology. *Oxford Surveys in Evolutionary Biology* **5**, 215–45.

Ewer, R. F. (1968). *Ethology of mammals*. Plenum, New York.

Fagen, R. (1981). *Animal play behavior*. Oxford University Press, New York.

Falconer, D. S. (1981). *Introduction to quantitative genetics*, 2nd edition. Longman, London.

Farcas, S. R. and Shorey, H. H. (1974). Mechanisms of orientation to a distant pheromone source. In *Pheromones* (ed. M. C. Birch), pp. 81–95, North-Holland, Amsterdam.

Felsenstein, J. (1985*a*). Recombination and sex: is Maynard Smith necessary? In *Evolution. Essays in honor of John Maynard Smith* (ed. P. J. Greenwood, P. H. Harvey, and M. Slatkin), pp. 209–19, Cambridge University Press.

Felsenstein, J. (1985*b*). Phylogenies and the comparative method. *American Naturalist* **125**, 1–15.

Felsenstein, J. (1988). Phylogenies and quantitative characters. *Annual Review of Ecology and Systematics* **19**, 445–71.

Fisher, R. A. (1930). *The genetical theory of natural selection*. Oxford University Press.

Fletcher, D. J. C. and Michener, C. D. (1987). *Kin recognition in animals*. Wiley-Interscience, New York.

Foster, S. A. (1992). Evolution of the reproductive behavior of the threespine stickleback. In *Evolutionary biology of the threespine stickleback* (ed. M. A. Bell and S. A. Foster), Oxford University Press (in press).

Francis, R. C. (1984). The effects of bidirectional selection for social dominance on agonistic behavior and sex ratios in the paradise fish (*Macropodus opercularis*). *Behaviour* **90**, 25–45.

Francis, R. C., Baumgartner, J. V., Havens, A. C. and Bell, M. A. (1986). Historical and ecological sources of variation among lake populations of threespine sticklebacks, *Gasterosteus aculeatus*, near Cook Inlet, Alaska. *Canadian Journal of Zoology* **64**, 2257–65.

Frank, S. A. (1986). Hierarchical selection theory and sex ratios. I. General solutions for structured populations. *Theoretical Population Biology* **29**, 312–42.

Frank, S. A. and Slatkin, M. (1990). Evolution in a variable environment. *American Naturalist* **136**, 244–60.

French, J. A. and Cleveland, J. (1984) Scent-marking in the tamarin, *Saguinus oedipus*: Sex differences and ontogeny. *Animal Behavior* **32**, 615–23.

Frisch, K. von (1938). Zur Psychologie des Fisch-Schwarmes. *Naturwissenschaften* **26**, 601–6.

Futuyma, D. J. (1986). *Evolutionary biology*. 2nd edition. Sinauer, Sunderland, MA.

Gabriel, W. (1987). Quantitative genetic models for parthenogenetic species. In *Population genetics and evolution* (ed. G. de Jong), pp. 73–82, Berlin, Springer-Verlag.

Gear, A. J. and Huntley, B. (1991). Rapid changes in the range limits of Scots pine 4000 years ago. *Science* **151**, 544–7.

Gee, J. H. R. and Giller, P. S. (1987). *Organization of communities, past and present*. Blackwell Scientific, Oxford.

Geliva, E. A. (1987). Meiotic drive in the sex chromosome system of the varying lemming, *Dicrostomyx torquatus* Pall. (Rodentia, Microtinae). *Heredity* **59**, 383–9.

Ghiselin, M. T. (1969). The evolution of hermaphroditism among animals. *Quarterly Review of Biology* **44**, 189–208.

Ghiselin, M. T. (1974a). *The economy of nature and the evolution of sex*. University of California Press, Berkeley.

Ghiselin, M. T. (1974b). A radical solution to the species problem. *Systematic Zoology* **23**, 536–44.

Gibbs, H. L. (1988) Heritability and selection on clutch size in Darwin's medium ground finches (*Geospiza fortis*). *Evolution* **42**, 750–62.

Gibbs, H. L. and Grant, P. R. (1987). Ecological consequences of an exceptionally strong el nino event on Darwin's finches. *Ecology* **68**, 1735–46.

Gibson, R. M. (1988). Explaining the peacock's tail. *Science* **242**, 1583.

Gibson, R. M. and Bradbury, J. W. (1986) In *Ecological Aspects of Social Evolution* (ed. D. I. Rubenstein and R. W. Wrangham), pp. 379–98, Princeton University Press.

Gingerich, P. D. (1977) Patterns of evolution in the mammalian fossil record. In *Patterns of evolution, as illustrated by the fossil record* (ed. A. Hallam), pp. 469–500, Elsevier, Amsterdam.

Gittleman, J. L. and Kot, M. (1991). Adaptation: Statistics and a null model for estimating phylogenetic effects. *Systematic Zoology,* **39**, 227–44.

Glass, B. (1985). Explanation in biology. In *Progress or catastrophe*, Convergence, A series founded and planned by R. N. Anshen. (ed. R. N. Anshen), pp. 33–64, Praeger, New York.

Gleick, J. (1987). *Chaos: making a new science*. Penguin Books, New York.

Goldschmidt, R. B. (1940). *The material basis of evolution*. Yale University Press, New Haven.

Goldsmith, T. H. (1990). Optimization, constraint, and history in the evolution of eyes. *Quarterly Review of Biology* **65**, 281–322.

Gonzalez, A. and Mensua, J. L. (1987). Genetic polymorphisms and high genetic load in natural populations of *Drosophila melanogaster* from cellar and vinyard. *Heredity* **59**, 227–36.

Goodnight, C. J. (1990). Experimental studies of community evolution. I: The response to selection at the community level. *Evolution* **44**, 1614–24.

Gould, J. L. (1982). *Ethology*. Norton, New York.

Gould, J. L. (1986). The locale map of honey bees: Do insects have cognitive maps? *Science* **232**, 861–3.

Gould, S. J. (1973). Positive allometry of antlers in the 'Irish Elk', *Megaloceros giganteus*. *Nature* **244**, 375–6.

Gould, S. J. (1980). The panda's thumb.

Gould, S. J. (1982*a*). Darwinism and the expansion of evolutionary theory. *Science* **216**, 380–7.

Gould, S. J. (1982*b*). The meaning of punctuated equilibrium and its role in validating a hierarchical approach to macroevolution. In *Perspectives on evolution* (ed. R. Milkman), pp. 83–104, Sinauer, Sunderland, MA.

Gould, S. J. (1983). Irrelevance, submission, and partnership: The changing role of palaeontology in Darwin's three centennials, and a modest proposal for macroevolution. In *Evolution from molecules to men* (ed. D.S. Bendall), pp. 347–66, Cambridge University Press.

Gould, S. J. (1989). *Wonderful life*. Norton, New York.

Gould, S. J. and Eldredge, N. (1977). Punctuated equilibria: the tempo and mode of evolution revisited. *Paleobiology* **6**, 115–151.

Gould, S. J. and Lewontin, R. C. (1979). The spandrels of San Marco and the Panglossian paradigm: a critique of the adaptationist program. *Proceedings of the Royal Society of London, B* **205**, 581–98.

Gould, S. J. and Vrba, E. S. (1982). Exaptation — a missing term in the science of form. *Paleobiology* **1**, 4–15.

Gouyon, P.-H. and Gliddon, C. (1988). The genetics of information and the evolution of avatars. In *Population genetics and evolution* (ed. G. de Jong), pp. 119–23, Springer-Verlag, Berlin.

Govind, C. K. (1989). Asymmetry in lobster claws. *American Scientist* **77**, 468–74.

Grafen, A. (1985). A geometric view of relatedness. *Oxford Surveys in Evolutionary Biology* **2**, 28–89.

Graham, R. W. (1986). Response of mammalian communities to environmental changes during the late Quaternary. In *Community ecology* (ed. J. Diamond and T. J. Case), pp. 300–13, Harper and Row, New York.

Grasse, P.-P. (1955). *Traite de Zoologie. Tome XVII. Mammiferes.* Masson et Cie Editeurs, Paris.

Grasse, P.-P. (1958). *Traite de Zoologie. Tome XIII.* Masson et Cie Editeurs, Paris.

Graul, W. D., Derrickson, S. R. and Mork, D. W. (1977). The evolution of avian polyandry. *American Naturalist* **111**, 812–6.

Greene, E. (1989). A diet-induced developmental polymorphism in a caterpillar. *Science* **243**, 643–6.

Griffin, D. (1981). *The question of animal awareness*, 2nd edition. W Kaufmann. Los Altos, CA.

Grosberg, R. K. (1988). Life-history variation within a population of the colonial ascidian *Botryllus schlosseri*. I. The genetic and environmental control of seasonal variation. *Evolution* **42**, 900–21.

Gross, M. R. (1985). Disruptive selection for alternative life histories in salmon. *Nature* **313**, 47–8.

Gross, M. R. (1987) Evolution of diadromy in fish. *American Fishery Society Symposium* **1**, 14–25.

Gross, M. R. and Charnov, E. L. (1980). Alternative male life histories in bluegill sunfish. *Proceedings of the National Academy of Sciences (U.S.)* **77**, 6937–40.

Gross, M. R. and MacMillan, A. M. (1981). Predation and the evolution of colonial nesting in bluegill sunfish (*Lepomis macrochirus*). *Behavioral Ecology and Sociobiology* **8**, 163–74.

Grudzien, T. A. and Turner, B. J. (1984). Direct evidence that the *Ilyodon* morphs are a single biological species. *Evolution* **38**, 402–7.

Gustafsson, L. (1986). Lifetime reproductive success and heritability: Empirical support for Fisher's fundamental theorem. *American Naturalist* **128**, 761–4.

Hagedorn, M. (1988). Ecology and behavior of a pulse-type electric fish, *Hypopomus occidentalis* (Gymnotiformes, Hypopomidae), in a fresh-water stream in Panama. *Copeia* 324–35.

Hairston, N. G., Jr. (1990). Fluctuating selection and response in a population of freshwater copepods. *Evolution* **44**, 1796–805.

Haldane, J. B. S. (1932). *The causes of evolution*. Longmans, London.

Haldane, J. B. S. (1937). The effect of variation on fitness. *American Naturalist* **71**, 337–49.

Haldane, J. B. S. (1957). The cost of natural selection. *Journal of Genetics* **55**, 511–24.

Hamilton, W. D. (1964). The genetical evolution of social behaviour. I and II. *Journal of Theoretical Biology* **7**, 1–52.

Hamilton, W. D. (1966). The moulding of senescence by natural selection. *Journal of Theoretical Biology* **12**, 12–45.

Hamilton, W. D., Axelrod, R. and Tanese, R. 1990. Sexual reproduction as an adaptation to resist parasites (A review). *Proceedings of the National Academy of Sciences (U.S.A.)* **87**, 3566–73.

Hammerstein, P. and Parker, G. A. (1987). Sexual selection: Games between the sexes. In *Sexual selection: Testing the alternatives*, Life Sciences Research Report 39 (ed. J. W. Bradbury and M. B. Andersson), pp. 119–42, Wiley, Chichester.

Harden Jones, F. M. (1968). *Fish migration*. Edward Arnold, London.

Harper, J. L. (1977). *The population biology of plants*. Academic Press, New York.

Harrington, R. W., Jr. (1975). Sex determination and differentiation among uniparental homozygotes of the hermaphroditic fish *Rivulus marmoratus* (Cyprinodontidae: Atheriniformes). In *Intersexuality in the animal kingdom* (ed. R. Reinboth), pp. 249–62, Springer-Verlag, Berlin.

Harrison, R. G. and Rand, D. M. (1989). Mosaic hybrid zones and the nature of species boundaries. In *Speciation and its consequences* (ed. D. Otte and J. A. Endler), pp. 111–33, Sinauer, Sunderlands, MA.

Harvell, C. D. (1990). The ecology and evolution of inducible defenses. *Quarterly Review of Biology* **65**, 323–40.

Harvey, P. H. and Pagel, M. D. (1991). *The comparative method in evolutionary biology*. Oxford University Press.

Harvey, P. H. and Partridge, L. (1987). Murderous mandibles and black holes in hymenopteran wasps. *Nature* **326**, 128–9.

Hawkes, K., O'Connell, J. F., Hill, K. and Charnov, E. L. (1985). How much is enough? Hunters and limited needs. *Ethology and Sociobiology* **6**, 3–15.

Hendrix, S. D. and Trapp, E. J. (1989). Floral herbivory in *Pastinaca sativa*: Do compensatory responses offset reduction in fitness. *Evolution* **43**, 891–5.

Hengeveld, R. (1988). Mayr's ecological species criterion. *Systematic Zoology* **37**, 47–55.

Hepper, P. G. (1986). Kin recognition: Functions and mechanisms: a review. *Biological Reviews* **61**, 63–93.

Hervey, G. F. and Hems, J. (1968). *The goldfish*. Faber and Faber, London.

Hickman, C. P. (1955). *Integrated principles of zoology*. Mosby, St Louis.

Hillis, D. M. (1988). Systematics of the *Rana pipiens* complex: Puzzle and paradigm. *Annual Review of Ecology and Systematics* **19**, 39–63.

Hines, W. G. S. (1987). Evolutionary stable strategies: A review of basic theory. *Theoretical Population Biology* **31**, 195–272.

Hoffman, A. (1986). Neutral model of Phanerozoic diversification: Implications for macroevolution. *Neues Jahrbuch fuer Geologie und Palaontologie, Abhandlung* **172**, 219–44.

Hoffman, A. (1987). Neutral model of taxonomic diversification in the Phanerozoic: A methodological discussion. *Neutral Models in Biology*. (ed. M. H. Nitecki and A. Hoffman), pp. 133–46, Oxford University Press.

Holman, E. W. (1987). Recognizability of sexual and asexual species of rotifers. *Systematic Zoology* **36**, 381–6.

Horan, B. L. (1989). Functional explanations in sociobiology. *Biology and Philosophy* **4**, 131–58.

Houde, A. E. (1988) Genetic differences in female choice between two guppy populations. *Animal Behavior* **36**, 510–16.

Hrdy, S. B. (1981) *The woman that never evolved*. Harvard University Press, Cambridge, MA.

Hubbs, C. L. and Hubbs, L. C. (1944). Bilateral asymmetry and bilateral variation in fishes. *Papers of the Michigan Academy of Science* **30**, 229–310.

Hubbs, C. L. and Whitlock, S. C. (1929). Diverse types of young in a single species of fish, the gizzard shad. *Papers of the Michigan Academy of Science, Arts and Letters* **10**, 461–82.

Huck, U. W., Seger, J. and Lisk, R. W. (1990). Litter sex ratios in the golden hamster vary with time of mating and with litter size and are not binomially distributed. *Behavioral Ecology and Sociobiology* **26**, 99–109.

Hughes, R. N. (1989). *Functional biology of clonal animals*. Chapman and Hall, New York.

Hull, D. L. (1976). Are species really individuals? *Systematic Zoology* **25**, 174–91.

Hull, D. L. (1978). A matter of individuality. *Philosophy of Science* **45**, 335–60.

Hull, D. L. (1988). Interactors versus vehicles. In *The role of behavior in evolution* (ed. H. C. Plotkin), pp.19–50, MIT Press, Cambridge, MA.

Humphrey, N. (1976). The social function of intellect. In *Growing points in ethology* (ed. P. P. G. Bateson and R. A. Hinde), pp. 303–17, Cambridge University Press.

Huxley, J. S. (1942). *Evolution, the modern synthesis*. Harper and Brothers, New York.

Huxley, J. S. (1953). *Evolution in action*. Harper and Brothers, New York.

Huxley, J. S. (1954). The evolutionary process. In *Evolution as a process* (ed. J. Huxley, A. C. Hardy and E. B. Ford), pp. 1–23, Allen and Unwin, London.

Innes, D. J. (1989). Genetics of *Daphnia obtusa*: genetic load and linkage analysis in a cyclical parthenogen. *Journal of Heredity* **80**, 6–10.

Iwamoto, R. N., Alexander, B. A. and Hershberger, W. K. (1984). Genotypic and environmental effects on the incidence of sexual precocity in coho salmon (*Oncorhynchus kisutch*). *Aquaculture* **43**, 105–21.

Jablonski, D. (1986a). Background and mass extinctions: The alternation of macroevolutionary regimes. *Science* **231**, 129–33.

Jablonski, D. (1986b). Evolutionary consequences of mass extinctions. In *Patterns and processes in the history of life*, Life Sciences Research Report no. 36 (ed. D. Jablonski and D. M. Raup), pp. 313–29, Springer-Verlag, Berlin.

Jacobson, M. (1972). *Insect sex pheromones*. Academic Press, New York.

James, F. C., Johnston, R. F., Wamer, N. O., Niemi, G. J. and Boecklen, W. J. (1984). The Grinellian niche of the wood thrush. *American Naturalist* **124**, 17–47.

Janzen, D.H. (1977). What are dandelions and aphids? *American Naturalist* **111**, 586–9.

Jeanne, R. L. (editor) (1988). *Interindividual behavioral variability in social insects*. Westview, Boulder.

Jensen, J. S. (1990). Plausibility and testability: Assessing the consequences of evolutionary innovation. In *Evolutionary innovations* (ed. M. H. Nitecki), pp. 171–90, University of Chicago Press.

Johannsen, W. (1909). *Elemente der exakten Erblichkeitslehre*. Gustaf Fischer, Jena.

John-Adler, H. B., Morin, P. J. and Lawler, P. (1988). Thermal physiology, phenology, and distribution of tree frogs. *American Naturalist* **132**, 506–20.

Johnson, H. A. (1987). Thermal noise and biological information. *Quarterly Review of Biology* **62**, 141–52.

Johnston, R. F. and Selander, R. K. (1964). House sparrows: rapid evolution of races in North America. *Science* **144**, 548–50.

Johnston, R. F. and Selander, R. K. (1971). Evolution in the house sparrow. II. Adaptive differentiation in North American populations. *Evolution* **25**, 1–28.

Johnston, R. F. and Selander, R. K. (1973). Evolution in the house sparrow. III. Variation in size and sexual dimorphism in Europe and North and South America. *American Naturalist* **107**, 373–90.

Kat, P. W. (1985). Historical evidence for fluctuation in levels of hybridization. *Evolution* **39**, 1164–9.

Kauffman, S. A. (1985). Self organization, selective adaptation and its limits: A new pattern of inference in evolution and development. In *The new biology and the new philosophy of science* (ed. D. J. DePew and B. H. Weber), pp. 169–207, MIT Press, Cambridge, MA.

Kauffman, S. A. (1987). Self-organization, selective adaptation, and its limits: A new pattern of inference in evolution and development. In *Neutral models in biology* (ed. M. H. Nitecki and A. Hoffman), pp. 56–89, Oxford University Press.

Kingsley, M. C. S. and Ramsay, M. A. (1988). The spiral tusk of the narwhal. *Arctic* **41**, 236–8.

Kirkpatrick, M. (1985). Evolution of female choice and male parental investment in polygynous species: The demise of the 'sexy son.' *American Naturalist* **125**, 788–810.

Kirkpatrick, M. (1987). The evolutionary forces acting on female mating preferences in polygynous animals. In *Sexual selection: testing the alternatives*, Life Sciences Research Report no. 39 (ed. J.W. Bradbury and M. B. Andersson), pp. 67–82, Chichester, Wiley.

Klomp, H. F (1970). The determination of clutch size in birds: a review. *Ardea* **58**, 1–124.

Knight, C. A. and DeVries, A. L. (1989). Melting inhibition and superheating of ice by an antifreeze glycopeptide. *Science* **245**, 505–7.

Knowler, W.C, Pettitt, D. J., Bennett, P. H. and Williams, R. C. (1983). Diabetes mellitus in the Pima Indians: Genetic and evolutionary considerations. *American Journal of Physical Anthropology* **62**, 107–14.

Koehn, R. K. and Hilbish, T. J. (1987). The adaptive importance of genetic variation. *American Scientist* **75**, 134–41.

Kolman, W. A. (1960). The mechanism of natural selection for the sex ratio. *American Naturalist* **94**, 373–77.

Korpelainen, H. (1990). Sex ratios and conditions required for environmental sex determination in animals. *Biological Reviews (Cambridge)* **65**, 147–84.

Krasnoff, S. B. and Roelofs, W. L. (1988). Sex pheromone released as an aerosol by the moth *Pyrrharctia isabella*. *Nature* **333**, 263–5.

Krebs, J. R. and Davies, N. B. (editors) (1984). *Behavioral ecology. An evolutionary approach*. 2nd edition, Sinauer, Sunderland, MA.

Krebs, J. R. and Davies, N. B. (1987). *An introduction to behavioural ecology*, 2nd edition, Blackwell, Oxford.

Krebs, J. R. and Dawkins, R. (1984). Animal signals: Mind-reading and manipulation. In *Behavioral ecology. An evolutionary approach*, 2nd edition (ed. J. R. Krebs and N. B. Davies), pp. 380–402, Sinauer, Sunderland, MA.

Lack, D. (1954*a*). *The natural regulation of animal numbers*. Oxford University Press.

Lack, D. (1954*b*). The evolution of reproductive rates. In *Evolution as a process* (ed. J. Huxley, A. C. Hardy and E. B. Ford), pp. 143–56, Allen and Unwin, London.

Lakatos, I. (1978). Falsification and the methodology of scientific research programmes. In *The methodology of scientific research programmes* (I. Lakatos: Philosophical Papers, Vol. 1), pp. 8–10, Cambridge University Press.

Lambert, D. M. and Paterson, E. (1982). Morphological resemblance and its relationship to genetic distance measures. *Evolutionary Theory* **5**, 291–300.

Lande, R. (1975). The maintenance of genetic variation by mutation in a polygenic character with linked loci. vs selection. *Genetical Research* **26**, 221–35.

Lande, R. (1976). Natural selection and random genetic drift in phenotypic evolution. *Evolution* **30**, 314–34.

Lande, R. (1987). Genetic correlations between the sexes in the evolution of sexual dimorphism and mating preferences. In *Sexual selection: testing the alternatives* (ed. J. W. Bradbury and M. B. Andersson), pp. 83–94, Wiley, Chichester.

Langton, C. G. (1989). Artificial life. In *Artificial Life* (ed. C. G. Langton), pp. 1–24, Addison-Wesley, Redwood City.

Larson, A. (1984). Neontological inferences of evolutionary pattern and process in the salamander family Plethodontidae. *Evolutionary Biology* **17**, 119–217.

Lavin, P. A. and McPhail, J. D. (1987). Morphological divergence and the organization of trophic characters among lacustrine populations of the threespine stickleback (*Gasterosteus aculeatus*). *Canadian Journal of Fisheries and Aquatic Science* **44**, 1820–9.

Lawrey, J. D. (1984). *Biology of lichenized fungi*. Praeger, New York.

Layzer, J.B. and Clady, M. D. (1987). Phenotypic variation of young-of-the-year bluegills (*Lepomis macrochirus*) among microhabitats. *Copeia* (3), 702–7.

Leigh, E. G., Jr. (1977). How does selection reconcile individual advantage with the good of the group? *Proceedings of the National Academy of Sciences (U.S.A.)* **74**, 4542–6.

Lenski, R. E. (1988). Experimental studies of pleiotropy and epistasis in *Escherichia coli*. I. Variation in competitive fitness among mutants resistant to virus T4. *Evolution* **42**, 425–32.

Lerner, I. M. (1953). *Genetic homeostasis*. Wiley, New York.

Lessios, H. A. and Cunningham, C. W. (1990). Genetic incompatibility between species of the sea urchin *Echinometra* on the two sides of the Isthmus of Panama. *Evolution* **44**, 933–41.

Levin, L. A., Zhu, J. and Creed, E. (1991). The genetic basis of life-history characters in a polychaete exhibiting planktotrophy and lecithotrophy. *Evolution* **45**, 380–97.

Levins, R. (1968). *Evolution in changing environments*. Princeton University Press.

Levinton, J. S. (1986). Punctuated equilibrium. *Science* **231**, 1490.

Levinton, J. S. (1987). *Genetics, paleontology, and macroevolution*. Cambridge University Press, New York.

Lewis, T. (1984). Elements and frontiers of insect communication. In *Insect communication* (ed. T. Lewis), pp. 1–27, Academic Press, Orlando.

Lewontin, R. C. (1970). The units of selection. *Annual Review of Ecology and Systematics* **1**, 1–18.

Lewontin, R. C. and Hubby, J. L. (1966). A molecular approach to the study of heterozygosity in natural populations. II. Amount of variation and degree of heterozygosity in natural populations of *Drosophila pseudoobscura*. *Genetics* **54**, 595–609.

Liem, K. F. (1990). Key evolutionary innovations, differential diversity, and synecomorphosis. In *Evolutionary innovations* (ed. M. H. Nitecki), pp. 147–70, University of Chicago Press.

Liem, K. F. and Kaufman, L. (1984). Intraspecific macroevolution: Functional biology of the polymorphic cichlid species *Cichlasoma minckleyi*. In *Evolution of fish species flocks* (ed. A. A. Echelle and I. Kornfield), pp. 203–15, University of Maine Press, Orono.

Liles, G. (1988). Why is life so complex? *MBL Science* **3**, 9–13.

Linn, C. E., Jr., Campbell, M. G. and Roelofs, L. W. L. (1987). Pheromone components and active spaces: What do moths smell and where do they smell it? *Science* **237**, 650–2.

Liu, K.-B. (1990). Holocene paleoecology of the boreal forest in northern Ontario. *Ecological Monographs* **60**, 179–212.

Livingston, R. J. (1982). Trophic organization of fishes in a coastal seagrass system. *Marine Ecology Progress Series* **7**, 1–12.

Livingston, R. J. (1988). Inadequacy of species-level designations for ecological studies of coastal migratory fishes. *Environmental Biology of Fishes* **22**, 225–34.

Lloyd, E. A. (1988). *The structure and confirmation of evolutionary theory*, Contributions in Philosophy, no. 37, Greenwood Press, New York.

Loeschcke, V. (1987). *Genetic constraints on adaptive evolution*. Springer-Verlag, New York.

Lumsden, C. and Wilson, E. O. (1981). *Genes, mind, and culture: The coevolutionary process*. Harvard University Press, Cambridge, MA.

Lynch, C. B., Sulzbach, D. S. and Connolly, M. S. (1988). Quantitative genetic analysis of temperature regulation in *Mus domesticus*. IV. Pleiotropy and genotype-by-environment interactions. *American Naturalist* **132**, 521–37.

Lynch, M. (1988). The rate of polygenic mutation. *Genetical Research* **51**, 137–48.

Lynch, M. (1990). Mutation load and the survival of small populations. *Evolution* **44**, 1725–37.

McArthur, A. J. and Clark, J. A. (1988). Body temperature of homeotherms and the conservation of energy and water. *Journal of Thermal Biology* **13**, 9–13.

McClintock, B. (1965). The control of gene action in maise. *Brookhaven Symposium in Biology* **18**, 162–84.

McClintock, B. (1987). Introduction. In *Genes, cells, and organisms: Great books in experimental biology* (ed. J. A. Moore), pp. vii–xi, Garland, New York.

McDonald, D. B. (1989). Cooperation under sexual selection: Age-graded changes in a lekking bird. *American Naturalist* **134**, 709–30.

Mackie, G. O. (1986). From aggregates to integrates: physiological aspects of modularity in colonial animals. *Philosophical Transactions of the Royal Society of London, B* **313**, 175–96.

McNaughton, S. J. (1983). Compensatory plant growth as a response to herbivory. *Oikos* **40**, 329–36.

McNaughton, S. J. (1986). On plants and herbivores. *American Naturalist* **128**, 765–70.

Markl, H. (1972). Aggression und Beuteverhalten bei Piranhas (Serrasalminae, Characidae). *Zeitschrift fuer Tierpsychologie* **30**, 190–216.

Marliave, J. B. (1989). Variation in pigment and nape morphology of larval tidepool sculpins. In *11th Annual Larval Fish Conference*, American Fisheries Society Symposium 5 (ed.R. D. Hoyt), pp. 80–88, American Fisheries Society, Bethesda.

Marshall, L. G., Webb, S. D., Sepkoski, J. J.,Jr. and Raup, D. M. (1982). Mammalian evolution and the Great American Interchange. *Science* **215**, 1351–7.

Maschinski, J. and Whitham, T. G. (1989). The continuum of plant responses to herbivory: The influence of plant association, nutrient availability, and timing. *American Naturalist* **134**, 1–19.

Mason, R. T., Fales, H. M., Jones, T. H., Pannell, L. K., Chinn, J. W. and Crews, D. (1989). Sex pheromones in snakes. *Science* **245**, 290–3.

Maugh, T. H., II (1981). Biochemical markers identify mental states. *Science* **114**, 39–41.

May, M. T. (translator) (1968). *Galen on the usefulness of the parts of the body.* Cornell University Press, Ithaca.

Maynard Smith, J. (1964). Group selection and kin selection. *Nature* **201**, 1145–7.

Maynard Smith, J. (1974). The theory of games and the evolution of animal conflicts. *Journal of Theoretical Biology* **47**, 209–21.

Maynard Smith, J. (1976). Group selection. *Quarterly Review of Biology* **51**, 277–83.

Maynard Smith, J. (1978). *The evolution of sex*. Cambridge University Press, London.

Maynard Smith, J. (1982). *Evolution and the theory of games*. Cambridge University Press.

Maynard Smith, J. (1984*a*). The population as a unit of selection. In *Evolutionary ecology*, 23rd British Ecological Society Symposium (ed. B. Shorrocks), pp. 195–202, Blackwell, Oxford.

Maynard Smith, J. (1984*b*). The ecology of sex. In *Behavioural ecology: an evolutionary approach* (J. R. Krebs and N. B. Davies), pp. 201–21, Sinauer, Sunderland, MA.

Maynard Smith, J. (1988*a*) The evolution of recombination. In *The evolution of sex* (ed. R. E. Michod and B. R. Levin), pp. 106–25, Sinauer, Sunderland, MA.

Maynard Smith, J. (1988*b*). *Did Darwin get it right?* Chapman and Hall, New York.

Maynard Smith, J. (1989). *Evolutionary genetics*. Oxford University Press.

Maynard Smith, J. and Sondhi, K. C. (1960). The genetics of a pattern. *Genetics* **45**, 1039–50.

Maynard Smith, J. , Burian, R., Kauffman, S., Alberch, P., Campbell, J., Goodwin, B., Lande, R., Raup, D. and Wolpert, L. (1985). Developmental constraints and evolution. *Quarterly Review of Biology* **60**, 265–87.

Mayr, E. (1942). *Systematics and the origin of species from the viewpoint of a zoologist*. Columbia University Press, New York.

Mayr, E. (1954). Change of genetic environment and evolution. In *Evolution as a process* (ed. J. Huxley, A. C. Hardy and E. B. Ford), pp. 157–80, Macmillan, New York.

Mayr, E. (1966). *Animal species and evolution*. Harvard University Press, Cambridge, MA.

Mayr, E. (1976). *Evolution and the diversity of life*. Harvard University Press, Cambridge, MA.

Mayr, E. (1982). *The growth of biological thought: Diversity, evolution, and inheritance*. Belknap Press, Cambridge, MA.

Mayr, E. (1983). How to carry out the adaptationist program? *American Naturalist* **121**, 324–34.

Mayr, E. (1988). *Toward a new philosophy of biology*. Harvard University Press, Cambridge, MA.

Meglitch, P. A. (1967). *Invertebrate zoology*. Oxford University Press, New York.

Meyer, A. (1987). Phenotypic plasticity and heterochrony in *Cichlasoma managuense* (Pisces, Cichlidae) and their implications for speciation in cichlid fishes. *Evolution* **41**, 1357–69.

Meyer, A. (1990). Morphometrics and allometry in the trophically polymorphic cichlid fish, *Cichlasoma citrinellum*: Alternative adaptations and ontogenetic changes in shape. *Journal of Zoology, London* **221**, 237–260.

Michod, R. E. (1982). The theory of kin selection. *Annual Review of Ecology and Systematics* **13**, 23–55.

Michod, R. E. and Levin, B. R. (ed.) (1988). *The evolution of sex*. Sinauer, Sunderland, MA.

Midgley, M. (1985). *Evolution as religion. Strange hopes and stranger fears*. Methuen, New York.

Milkman, R. (1982). Toward a unified selection theory. In *Perspectives on evolution* (ed. R. Milkman), pp. 105–18, Sinauer, Sunderland, MA.

Miller, J. A. (1990). Genes that protect against cancer. *BioScience* **40**, 563–6.

Mitchell, W. A. and Valone, T. J. (1990). The optimization research program: Studying adaptations by their function. *Quarterly Review of Biology* **65**, 43–52.

Mitchell-Olds, T. and Shaw, R. G. (1990). Comments on the causes of natural selection. *Evolution* **44**, 2158.

Mittelbach, G. (1984). Group size and feeding rate in bluegills. *Copeia* 998–1000.

Mitter, C., Farrell, B. and Wiegmann, B. (1988). The phylogenetic study of adaptive zones: Has phytophagy promoted insect diversification? *American Naturalist* **132**, 107–28.

Monahan, R. K. (1988). *Sex ratios and sex change in Ophryotrocha puerilla puerillis (Polychaeta, dorvalleidae)*. Ph.D. thesis. State University of New York at Stony Brook.

Moore, A. J. and Moore, P. J. (1988). Female strategy during mate choice: Threshold assessment. *Evolution* **42**, 387–91.

Moorhead, P. S. and Kaplan, M. M. (1985, reprint of 1967 edition) *Mathematical challenges to the neo-Darwinian interpretation of evolution*. Alan R. Liss, New York.

Mori, S. (1990). Two morphological types in the reproductive stock of the three-spined stickleback, *Gasterosteus aculeatus*, in Lake Harutori, Hokkaido Island. *Environmental Biology of Fishes* **27**, 21–31.

Morton, E. S. (1990). Habitat segregation by sex in the hooded warbler: Experiments on proximate causation and discussion of its evolution. *American Naturalist* **135**, 319–33.

Mosterin, J. (1986). Cultura como informacin. In *La sociedad naturalizada, genetica y conducta* (ed. J. Sanmartin), pp. 351–67, Tiratn lo Blanch, Valencia.

Mosterin, J. (1988). Ontological queries and evolutionary processes. Comments on Hull. *Biology and Philosophy* **3**, 204–9.

Mousseau, T. A. (1987). Natural selection and the heritability of fitness components. *Heredity* **59**, 181–97.

Muller, H. J. (1948). Evidence of the precision of genetic adaptation. *Harvey Lectures* **43**, 165–229.

Muller, H. J. (1964). The relation of recombination to mutational advance. *Mutation Research* **1**, 2–9.

Nap, J.-P. and Bisseling, T. (1990). Developmental biology of a plant-prokaryote symbiosis: The legume root nodule. *Science* **250**, 948–54.

Nesse, R. M. (1988). Life table tests of evolutionary theories of senescence. *Experimental Gerontology* **23**, 445–53.

Nevo, E. (1986). Mechanisms of adaptive speciation at the molecular and organismal levels. In *Evolutionary Processes and Theory* (ed. S. Karlin and E. Nevo), pp. 439–74, Academic Press, Orlando.

Noakes, D. L. G., Skulason, S. and Snorrason, S. S. (1989). Alternative life-history styles in salmonine fishes with emphasis on arctic char, *Salvelinus alpinus*. *Perspectives in Vertebrate Science* **6**, 329–46.

Noirot, C. and Pasteels, J. M. (1988). The worker caste is polyphyletic in termites. *Sociobiology* **14**, 15–20.

Nonacs, P. and Dill, L. (1990). Mortality risk vs. food quality trade-offs in a common currency: Ant patch preferences. *Ecology* **71**, 1886–92.

Noirot, C. (1989). Social structure in termite societies. *Ethology, Ecology, and Evolution* **1**, 1–17.

Nunney, L. (1985a). Group selection, altrusim, and structured-deme models. *American Naturalist* **126**, 212–30.

Nunney, L. (1985b). Female-biased sex ratios: Individual or group selection? *Evolution* **39**, 349–61.

Nunney, L. (1989). The maintenance of sex by group selection. *Evolution* **43**, 245–57.

Nunney, L. and Luck, R. F. (1988). Factors influencing the optimum sex ratio in a structured population. *Theoretical Population Biology* **33**, 1–30.

O'Hara, R. J. (1988). Homage to Clio, or, toward an historical phlosophy for evolutionary biology. *Systematic Zoology* **37**, 142–55.

Orzak, S. H. (1985). Population dynamics in variable environments. 5. The genetics of homeostasis revisited. *American Naturalist* **125**, 550–72.

Orzak, S. H. and Tuljapurkar, S. D. (1989). Population dynamics in variable environments. VII. The demography and evolution of iteroparity. *American Naturalist* **133**, 901–23.

Otte, D. (1974). Effects and functions in the evolution of signaling systems. *Annual Review of Ecology and Systematics* **5**, 385–417.

Otte, D. (1989). Speciation in hawaiian crickets. In *Speciation and its consequences* (ed. D. Otte and J. A. Endler), pp. 482–526, Sinauer, Sunderland, MA.

Otte, D. and Endler, J. A. (editors) (1989). *Speciation and its consequences* Sinauer, Sunderland, MA.

Pagel, M. D. and Harvey, P. H. (1988). Recent developments in the analysis of comparative data. *Quarterly Review of Biology* **63**, 413–40.

Paige, K. N. and Whitham, T. G. (1987). Overcompensation in response to mammalian herbivory: the advantage of being eaten. *American Naturalist* **129**, 407–16.

Paine, M. D. and Balon, E. K. (1986). Early development of johnny darter, *Etheostoma nigrum*, and the fantail darter, *E. flabellare*, with a discussion of its ecological and evolutionary aspects. *Environmental Biology of Fishes* **15**, 191–220.

Paley, William (1836). *Natural theology, Volume 1*. Charles Knight, London.

Paradis, J. and Williams, G. C. (1989). *Evolution and ethics*. Princeton University Press.

Parker, G. A. (1984). Evolutionarily stable strategies. In *Behavioural ecology: an evolutionary approach* (ed. J. R. Krebs and N. B. Davies), pp. 30–61, Sinauer, Sunderland, MA.

Parkin, D. T. (1987). Evolutionary genetics of house sparrows. In *Avian genetics* (ed. F. Cooke and P. A. Buckley), pp. 381–406, Academic Press, London.

Parks, A. L., Parr, B. A., Chin, J.-E., Leaf, D. S. and Raff, R. A. (1988). Molecular analysis of heterchronic changes in the evolution of direct developing sea urchins. *Journal of Evolutionary Biology* **1**, 27–44.

Parsons, P. A. (1982). Evolutionary ecology of Australian *Drosophila*: a species analysis. *Evolutionary Biology* **14**, 297–350.

Partridge, L. (1980). Mate choice increases a component of offspring fitness in fruit flies. *Nature* **283**, 290–1.

Payne, R. B. and Westneat, D. F. (1988). A genetic and behavioral analysis of mate choice and song neighborhoods in indigo buntings. *Evolution* **42**, 935–47.

Pease, C. M. and Bull, J. J. (1990). A critique of methods for measuring life-history trade-offs. *Journal of Evolutionary Biology* **1**, 293–303.

Pease, C. M., Lande, R. and Bull, J. J. (1989) A model of population growth, dispersal and evolution in a changing environment. *Ecology* **70**, 1657–64.

Peck, S. B. (1986). Evolution of adult morphology and life-history characters in cavernicolous *Ptomaphagus* beetles. *Evolution* **40**, 1021–30.

Pennak, R. W. (1978). *Freshwater invertebrates of the United States*, 2nd edition, Wiley, New York.

Pfeiffer, W. (1962). The fright reaction of fish. *Biological Reviews (Cambridge)* **37**, 495–511.

Pfeiffer, W. (1977) The distribution of fright reaction and alarm substance cells in fishes. *Copeia* 653–65.

Pfeiffer, W. (1982). Chemical signals in communication. In *Chemoreception in Fishes* (ed. T. J. Hara), pp. 307–26, Elsevier, Amsterdam.

Pimm, S. L., Jones, H. L. and Diamond, J. (1988). On the risk of extinction. *American Naturalist* **132**, 757–85.

Pinker, S. and Bloom, P. (1989). Natural language and natural selection. Center for Cognitive Science, Occasional Paper 39, 45 pp.

Pittendrigh, C. S. (1958). Adaptation, natural selection, and behavior. In *Behavior and Evolution* (ed. A. Roe and G. G. Simpson), pp. 390–416, Yale University Press, New Haven.

Polis, G. (1984). Age structure component of niche width and intraspecific resource partitioning: Can age groups function as ecological species? *American Naturalist* **123**, 541–64.

Power, M. E. (1990). Resource enhancement by indirect effects of grazers: Armored catfish, algae, and sediment. *Ecology* **71**, 897–904.

Pressley, P. H. (1981). Parental effort and the evolution of nest-guarding tactics in the threespine stickleback, *Gasterosteus aculeatus* L. *Evolution* **35**, 282–95.

Price, T. and Liou, L. (1989). Selection on clutch size in birds. *American Naturalist* **134**, 950–59.

Prosser, C. L. (1973). *Comparative animal physiology*, 3rd edition. Saunders, Philadalphia.

Prosser, C. L. (1986). *Adaptational biology*. Wiley, New York.

Protasov, V. R. (1966). Sonic signals of fishes. (In Russian with English summary.) *Referativnÿĭ Zhurnal Biologiya* **5I123**, 251–5.

Provine, W. B. (1986). *Sewall Wright and evolutionary biology*. University of Chicago Press.

Provine, W. B. (1989). Progress in evolution and meaning in life. In *Evolutionary progress* (ed. M. H. Nitecki), pp. 49–74, University of Chicago Press.

Pugesek, B. H. (1990). Parental effort in the California gull: Tests of parent–offspring conflict theory. *Behavioral Ecology and Sociobiology* **27**, 211–5.

Pulliam, H. R. and Caraco, T. (1984). Living in groups: Is there an optimum group size? In *Behavioural ecology: an evolutionary approach* (ed. J. R. Krebs and N. B. Davies), pp. 122–47, Sinauer, Sunderland, MA.

Queller, D. C. (1989). Inclusive fitness in a nutshell. *Oxford Surveys in Evolutionary Biology* **6**, 73–109.

Raikow, R. J. (1986). Why are there so many kinds of passerine birds? *Systematic Zoology* **35**, 255–9.

Ramsay, M. A. and Dunbrack, R. L. (1986). Physiological constraints on life history phenomena: The example of small bear cubs at birth. *American Naturalist* **127**, 735–43.

Rasa, A. E., Vogen, C. and Voland, E. (1989). *The sociobiology of sexual and reproductive strategies*. Chapman and Hall, New York.

Ratnieks, F. L. W. (1988). Reproductive harmony via mutual policing by workers in eusocial Hymenoptera. *American Naturalist* **132**, 217–36.

Raup, D. M. (1986). Biological extinction in earth history. *Science* **231**, 1528–33.

Raup, D. M. (1987). Neutral models in paleobiology. In *Neutral models in biology* (ed. M. H. Nitecki and A. Hoffman), pp. 121–32, Oxford University Press, New York.

Raven, P. H. (1986). Modern aspects of the biological species in plants. In *Modern aspects of species* (ed. K. Iwatsuki, P. H. Raven and W. J. Bock), pp. 11–29, University of Tokyo Press.

Reeve, R. E., Smith, E. and Wallace, B. (1988). Components of fitness become effectively neutral in equilibrium populations. *Proceedings of the National Academy of Sciences (U.S.A.)*, **87**, 2018–20.

Reznick, D. N. (1989). Life-history evolution in guppies: 2. Repeatability of field observations and the effects of season on life histories. *Evolution* **43**, 1285–97.

Reznick, D. N., Meyer, A. and Frear, D. (1992). Life history of *Brachyraphis rhabdophora* (Pisces: Poeciliidae). *Copeia* (in press).

Ribbink, A. J. (1984). The feeding behaviour of a cleaner and scale, skin, and fin eater from Lake Malawi (*Docimodus evelynae*: Pisces, Cichlidae). *Netherlands Journal of Zoology* **34**, 182–96.

Rice, S. H. (1990). A geometric model for the evolution of development. *Journal of Theoretical Biology* **143**, 319–42.

Ricklefs, R. E. and Marks, H. L. (1984). Insensitivity of brain growth to selection of four-week body mass in Japanese quail. *Evolution* **38**, 1180–5.

Ridley, M. (1983). *The explanation of organic diversity*. Oxford University Press.

Riget, F. F., Nygaard, K. H. and Christensen, B. (1986). Population structure, ecological segregation, and reproduction in a population of arctic char (*Salvelinus alpinus*) from Lake Taserquag, Greenland. *Canadian Journal of Fisheries and Aquatic Sciences* **43**, 985–92.

Rightmire, G. P. (1985). The tempo of change in the evolution of mid-Pleistocene *Homo*. In *Ancestors: The hard evidence* (ed. E. Delson), pp. 255–64, Alan R. Liss, New York.

Ritter, F. J. (1979). *Chemical ecology: Odour communication in animals*. Elsevier/North Holland, Amsterdam.

Roberts, L. (1989). How fast can trees migrate? *Science* **243**, 735–37.

Rockwell, R. F., Findlay, C. S. and Cooke, F. (1987). Is there an optimal clutch size in snow geese? *American Naturalist* **130**, 839–63.

Roelofs, W. L. and Carde, R. T. (1974). Sex pheromones in the reproductive isolation of lepidopterous species. In *Pheromones* (ed. M. C. Birch), pp. 96–114, North-Holland, Amsterdam.

Romer, A.S. and Grove, B. H. (1935). Environment of the early vertebrates. *American Midland Naturalist* **16**, 805–56.

Rose, M. R. (1991). *Evolutionary biology of aging*. Oxford University Press, New York, Oxford.

Rose, M. R., Service, P. M. and Hutchison, E. W. (1987) Three approaches to tradeoffs in life-history evolution. In *Genetic constraints on adaptive evolution* (ed. V. Loeschcke), pp. 91–105, Springer-Verlag, New York.

Roudebush, R. E. and Taylor, D. H. (1987). Chemical communication between two species of desmognathine salamanders. *Copeia* 744–8.

Roughgarden, J. and Pacala, S. (1989). Taxon cycle among *Anolis* lizard populations: Review of evidence. In *Speciation and its consequences* (ed. D. Otte and J. A. Endler), pp. 403–32, Sinauer, Sunderland, MA.

Rouhani, S. and Barton, N. (1987). Speciation and the 'shifting balance' in a continuous population. *Theoretical Population Biology* 31, 465–92.

Rowan, B. and Powers, D. A. (1991). A molecular genetic classification of zooxanthellae and the evolution of animal-algal symbioses. *Science* 251, 1348–51.

Rowland, W. J. (1989). Mate choice and the supernormality effect in female sticklebacks (*Gasterosteus aculeatus*). *Behavioral Ecology and Sociobiology* 24, 433–8.

Ruse, M. (1982). *Darwinism defended: A guide to the evolution controversies*. Addison-Wesley, Reading, MA.

Ruse, M. (1987). Biological species: Natural kinds, individuals, or what? *British Journal of Philosophy of Science* 38, 225–42.

Ruse, M. (1989) Is the theory of punctuated equilibrium a new paradigm? *Journal of Social and Biological Structure* 12, 195–212.

Ruud, J. T. (1965). The ice fish. *Scientific American* 213, 108–14.

Ryan, M.J. and Wagner, W. E., Jr. (1987). Asymmetries in mating preferences between species: Female swordtails prefer heterospecific males. *Science* 236, 595–7.

Ryder, O. A. (1986). Species conservation and systematics: the dilemma of subspecies. *Trends in Ecology and Evolution* 1, 9–10.

Sachs, T. (1988). Epigenetic selection: An alternative mechanism of pattern formation. *Journal of Theoretical Biology* 134, 547–59.

Sage, R. D. and Selander, R. K. (1975). Trophic radiation through polymorphism in cichlid fishes. *Proceedings of the National Academy of Sciences (U.S.A.)* 72, 4669–73.

Salthe, S. N. (1989). [Untitled.] In *Evolutionary biology at the crossroads* (ed. M. K. Hecht), pp. 174–6, Queens College Press, Flushing, NY.

Sargent, R. C. (1988). Paternal care and egg survival both increase with clutch size in the fathead minnow, *Pimephales promelas*. *Behavioral Ecology and Sociobiology* 23, 33–7.

Saunders, J. R. (1984) Antibiotic resistance in bacteria. *British Medical Bulletin* **40**, 54–60.

Schaffer, H. B. and Breden, F. (1989). The relationship between allozyme variation and life history: Non-transforming salamanders are less variable. *Copeia* (4), 1016–23.

Schmitt, J. and Ehrhardt, D. W. (1990). Enhancement of inbreeding depression by dominance and suppression in *Impatiens capensis. Evolution* **44**, 269–78.

Schneider, D. (1974). Sex attractant receptor of moths. *Scientific American* **231**, 28–35.

Schneider, D. (1978). Equalisation of prey numbers by migratory shore birds. *Nature* **271**, 353–4.

Seeley, T. D. (1989). The honey bee as a superorganism. *American Scientist* **77**, 546–53.

Semlitch, R. D., Harris, R. N. and Wilbur, H. M.(1990). Pedomorphosis in *Amblystoma talpoideum*, maintenance of population variation and alternative life-history pathways. *Evolution* **44**, 1604–13.

Setala, H. and Huhta, V. (1991). Soil fauna increase *Betula pendula* growth: Laboratory experiments with coniferous forest floor. *Ecology* **72**, 665–71.

Shallice, T. (1988). *From neuropsychology to mental structure*. Cambridge University Press, New York.

Shapiro, A. M. (1984). Polyphenism, phyletic evolution, and the structure of the pierid genome. *Journal of Research on the Lepidoptera* **23**, 177–95.

Shaw, E. (1978). Schooling fishes. *American Scientist* **66**, 166–75.

Sherman, P. W. (1988). Levels of analysis. *Animal Behavior* **36**, 616–9.

Sherman, P. W. (1989). The clitoris debate and levels of analysis. *Animal Behavior* **37**, 697–8.

Shine, R. (1985). The evolution of viviparity in reptiles: an ecological analysis. In *Biology of the reptilia* (ed. C. Gans and F. Billet), Vol. 15, pp. 605–94, Wiley, New York.

Shine, R. and Guillette, L. J., Jr. (1988). The evolution of viviparity in reptiles: A physiological model and its ecological consequences. *Journal of Theoretical Biology* **132**, 43–50.

Sibley, R. M. (1983). Optimal group size is unstable. *Animal Behavior* **31**, 947–8.

Simberloff, D. and Connor, E. F. (1981). Missing species combinations. *American Naturalist* **118**, 215–39.

Simmons, L. W. (1987). Female choice contributes to offspring fitness in the field cricket, *Gryllus bimaculatus* (De Geer). *Behavioral Ecology and Sociobiology* **21**, 313–21.

Simpson, G. G. (1944). *Tempo and mode in evolution*. Columbia University Press, New York.

Sinervo, B. and McEdward, L. R. (1988). Developmental consequences of an evolutionary change in egg size: An experimental test. *Evolution* **42**, 885–99.

Singer, P. (1981). *The expanding circle*. Farrar, Straus and Giroux, New York.

Slatkin, M. (1987). Gene flow and the geographic structure of natural populations. *Science* **236**, 787–92.

Slowinski, J. B. and Guyer, C. (1989). Testing null models in questions of evolutionary success. *Systematic Zoology* **38**, 189–91.

Smith, R. J. F. (1982). The adaptive significance of the alarm substance–fright reaction system. In *Chemoreception in fishes* (ed. T. J. Hara), pp. 327–42, Elsevier, Amsterdam.

Smith, R. J. F. and Smith, J. D. (1983). Seasonal loss of alarm substance cells in *Chrosomus neogaeus*, *Notropis venustus*, and *N. whipplei*. *Copeia* 822–6.

Smith, T. B. (1990). Resource use by bill morphs of an African finch: Evidence for intraspecific competition. *Ecology* **71**, 1246–57.

Smith, W. P. (1987). Maternal defense in Columbian white-tailed deer: When is it worth it? *American Naturalist* **130**, 310–6.

Smouse, P. E. and Li, W.-H. (1987). Likelihood analysis of mitochondrial restriction-cleavage patterns for the human-chimpanzee-gorilla trichotomy. *Evolution* **41**, 1162–76.

Sober, E. (1984). *The nature of selection*. The MIT Press, Cambridge, MA.

Sokal, R. R. and Crovello, T. J. (1970). The biological species: a critical evaluation. *American Naturalist* **104**, 127–53.

Solbrig, O. T. and Solbrig, D. J. (1984). Size inequalities and fitness in plant populations. *Oxford Surveys in Evolutionary Biology* **1**, 141–59.

Spiess, E. B. (1977). *Genes in populations*. Wiley, New York.

Stanley, S. M. (1979). *Macroevolution: Pattern and process*. W. H. Freeman, San Francisco.

Stearns, S. C. (1976). Life history tactics: a review of the ideas. *Quarterly Review of Biology* **51**, 3–47.

Stearns, S. C. (1983). The genetic basis of differences in life-history traits among six populations of mosquitofish (*Gambusia affinis*) that shared ancestors in 1905. *Evolution* **37**, 618–27.

Stearns, S. C. (1986). Natural selection and fitness, adaptation and constraint. In *Patterns and processes in the history of life: report of the Dahlem workshop* (ed. D. Jablonski and D. M. Raup), pp. 23–44, Springer-Verlag, Berlin.

Stearns, S. C. (editor) (1987). *The evolution of sex and its consequences*. Birkhauser-Verlag, Basel.

Stebbins, G. L. (1950). *Variation and evolution in plants*. Columbia University Press, New York.

Stenseth, N. C. (1985). Darwinian evolution in ecosystems: the Red Queen view. In *Evolution. Essays in honor of John Maynard Smith* (ed. P. H. Greenwood and P. H. Harvey), pp. 55–72, Cambridge University Press.

Sterelny, K. and Kitcher, P. (1988). The return of the gene. *Journal of Philosophy* **85**, 339–61.

Stiassny, M. L. J. and Jensen, J. S. (1987). Labroid interrelationships revisited. Morphological complexity, key innovations, and the study of comparative diversity. *Bulletin of the Museum of Comparative Zoology* **151**, 269–313.

Stone, A. R. and Hawksworth, D. L. (1987). *Coevolution and systematics*, Systematics Association Special Series, Vol. 32, Oxford University Press, New York.

Stoner, G. and Breeden, F. (1988). Phenotypic differentiation in female preference related to geographic variation in male predation risk in the Trinidad guppy (*Poecilia reticulata*). *Behavioral Ecology and Sociobiology* **22**, 285–91.

Storer, T. I. and Usinger, R. L. (1965). *General Zoology*. McGraw-Hill, New York.

Strathmann, R. R. (1978). The evolution and loss of feeding larval stages of marine invertebrates. *Evolution* **32**, 894–906.

Stumpke, H. (1957). *Bau und Leben der Rhinogradentia*. Gustaf Fischer Verlag, Stuttgart.

Sturtevant, A. H. (1938). Essays on evolution. II. On the effects of selection on social insects. *Quarterly Review of Biology* **13**, 74–6.

Suomalainen, E., Saura, A. and Lokki, J. (1987). *Cytology and evolution in parthenogenesis*. CRC Press, Boca Raton, FL.

Symons, D. (1979). *The evolution of human sexuality*. Oxford University Press.

Symons, D. (1986). Sociobiology and Darwinism. *Behavior and Brain Science* **9**, 208–9.

Symons, D. (1987). If we're all Darwinians, what's all the fuss about? In *Sociobiology and psychology* (ed. C. Crawford, D. Krebs and M. Smith), pp. 121–46, Lawrence Erlbaum, Hillsdale, NJ.

Symons, D. (1989). A critique of Darwinian anthropology. *Ethology and Sociobiology* **10**, 131–44.

Taussig, H. B. (1988). Evolutionary origin of cardiac malformations. *Journal of the American College of Cardiology* **12**, 1079–86.

Taylor, C. E., Perida, A. D. and Ferrari, J. A. (1987). On the correlation between mating success and offspring quality in *Drosophila melanogaster*. *American Naturalist* **129**, 721–9.

Taylor, P. D. and Sauer, A. (1980). The selective advantage of sex-ratio homeostasis. *American Naturalist* **116**, 305–10.

Taylor, P. D. and Williams, G. C. (1982). The lek paradox is not resolved. *Theoretical Population Biology* **22**, 392–409.

Taylor, P. D. and Williams, G. C. (1984). Demographic parameters at evolutionary equilibrium. *Canadian Journal of Zoology* **62**, 2264–71.

Taylor, P. D. and Wilson, D. W. (1988). A mathematical model for altruism in haystacks. *Evolution* **42**, 193–6.

Templeton, A. R. (1989). The meaning of species and speciation: A genetic perspective. In *Speciation and its consequences* (ed. D. Otte and J. A. Endler), pp. 3–27, Sinauer, Sunderland, MA.

Thomson, K. S. (1988). Marginalia: Ontogeny and phylogeny recapitulated. *American Scientist* **76**, 273–5.

Thornhill, R. (1989). Nest defense by red jungle fowl (*Gallus gallus spadiceus*) hens: The roles of renesting potential, parental experience and brood reproductive value. *Ethology* **83**, 31–42.

Thornhill, R. (1990). The study of adaptation. In *Interpretation and explanation in the study of behavior*, 2 vols (ed. M. Bekoff and D. Jamieson), pp. 1–31 , Westview Press, Boulder.

Tilman, D. (1982). *Resource competition and community structure*. Princeton University Press.

Tinbergen, N. (1965). Behavior and natural selection. *Ideas in modern biology* (ed. J. A. Moore), pp. 521–39. Natural History Press, Garden City, NY.

Turner, J. R. G. (1977). Butterfly mimicry: the genetical evolution of adaptation. *Evolutionary Biology* **10**, 163–206.

Turner, J. R. G. (1978). Why male butterflies are non-mimetic: natural selection, sexual selection, group selection, modification and sieving. *Biological Journal of the Linnean Society* **10**, 385–432.

Turner, J. R. G. (1984). Mimicry: The palatability spectrum and its consequences. In *The biology of butterflies* (ed. R. A. Vane-Wright and P. R. Ackery), pp. 141–61, Academic Press, London.

Turner, J. R. G. (1988). The evolution of mimicry: A solution to the problem of punctuated equilibrium. *American Naturalist* **131**, S42–66.

Underwood, A. J. (1986). The analysis of competition by field experiments. In *Community ecology: Pattern and process* (ed. D. J. Anderson and J. Kikkawa), pp. 240–68, Blackwell, Oxford.

Unger, L. M. (1983). Nest defense by deceit in the fathead minnow *Pimephales promelas*. *Behavioral Ecology and Sociobiology* **13**, 125–30.

Vadasz, C., Kobor, G. and Lajtha, A. (1983). Genetic dissection of a mammalian behaviour pattern. *Animal Behavior* **31**, 1029–36.

Valentine, J. W. (1986). Fossil record of the origin of Baupläne and its implications. In *Patterns and processes in the history of life*, Life Sciences Research Report, 39) (ed. D. M. Raup and D. Jablonski), pp. 209–22, Springer-Verlag, Berlin.

Van Valen, L. M. (1975). Group selection, sex, and fossils. *Evolution* **29**, 87–94.

Van Valen, L. M. (1982). Integration of species. Stasis and biogeography. *Evolutionary Theory* **6**, 99–112.

Van Valen, L. M. (1988). Species, sets, and the derivative nature of philosophy. *Biology and Philosophy* **3**, 49–66.

Van Valen, L. M. (1989). Three paradigms of evolution. *Evolutionary Theory* **9**, 1–17.

Van Valen, L. M. and Maiorani, V. C. (1985). Patterns of origination. *Evolutionary Theory* **7**, 107–25.

Van Valen, L. M. and Mallin, G. W. (1967). Selection in natural populations. 7. New York babies (fetal life study). *Annals of Human Genetics* **31**, 109–27.

Van Vorhees, D. A. (1988). Sexual selection and the evolution of sex-differences in age and size at maturity in the platyfish, *Xiphophorus maculatus*. Ph.D. dissertation. State University of New York, Stony Brook.

Verheijen, F. J. and Reuter, J. H. (1969) The effect of alarm substance on predation among cyprinids. *Animal Behavior* **17**, 551–4.

Vermeij, G. J. (1987). *Evolution and escalation*. Princeton University Press.

Verner, J. (1965). Selection for sex ratio. *American Naturalist* **99**, 419–21.

Via, S. (1987). Genetic constraints on the evolution of phenotypic plasticity. In *Genetic constraints on adaptive evolution* (ed. V. Loeschcke), pp. 47–71, Springer-Verlag, New York.

Villee, C. A. (1954). *Biology*. 2nd edition, Saunders, Philadelphia.

Vitt, L. J. (1986). Reproductive tactics of sympatric gekkonid lizards with a comment on the evolutionary and ecological consequences of invariant clutch size. *Copeia* (3), 773–86.

Vrba, E. S. (1984). What is species selection? *Systematic Zoology* **33**, 318–28.

Vrba, E. S. (1989). Levels of selection and sorting with special reference to the species level. *Oxford Surveys in Evolutionary Biology* **6**, 111–68.

Waddington, C. H. (1953). Genetic assimilation of an acquired character. *Evolution* **7**, 118–26.

Waddington, C. H. and Lewontin, R. C. (1968). A note on evolution and changes in the quantity of genetic information. In *Towards a theoretical biology* (ed. C. H. Waddington), Vol. 1, pp. 109–10, Edinburgh University Press.

Wade, M. J. (1978). A critical review of the models of group selection. *Quarterly Review of Biology* **53**, 101–14.

Wade, M. J. (1987). Measuring sexual selection. In *Sexual selection: testing the alternatives* (ed. J. W. Bradbury and M. B. Andersson), pp. 197–207, Wiley, Chichester.

Wade, M. J. and Breden, F. (1981). Effect of inbreeding on the evolution of altruistic behavior by kin selection. *Evolution* **35**, 844–58.

Wade, M. J. and Kalisz, S. (1990). The causes of natural selection. *Evolution* **44**, 1947–55.

Wagner, G. P. (1987). The significance of developmental constraints for phenotypic evolution by natural selection. In *Population genetics and evolution* (ed. G. de Jong), pp. 222–9, Springer-Verlag, Berlin.

Wagner, G. P. (1988). The influence of variation and of developmental constraints on the rate of multivariate phenotypic evolution. *Journal of Evolutionary Biology* **1**, 45–66.

Wagner, G. P. (1989). The origin of morphological characters and the biological basis of homology. *Evolution* **43**, 1157–71.

Wake, D.B., Roth, F. and Wake, M. H. (1983). On the problem of stasis in organismal evolution. *Journal of Theoretical Biology* **101**, 211–24.

Wake, D.B., Yanev, K. P. and Frelow, M. M. (1989). Sympatry and hybridization in a 'ring species': the plethodontid salamander *Ensatina eschscholtzii*. In *Speciation and its consequences* (ed. D. Otte and J. A. Endler), pp. 134–57, Sinauer, Sunderland, MA.

Waldman, B. (1982). Quantitative and developmental analysis of the alarm reaction in the zebra danio, *Brachydanio rerio*. *Animal Behavior* **30**, 1–9.

Waldman, B. (1988). The ecology of kin recognition. *Annual Review of Ecology and Systematics* **19**, 543–71.

Walker, E. and Paradiso, J. L. (1975). *Mammals of the world*, 3rd edition. Johns Hopkins University Press, Baltimore.

Wallace, B. (1970). *Genetic load: Its biological and conceptual aspects*. Prentice-Hall, Englewood Cliffs, NJ.

Wallace, B. (1987). Fifty years of genetic load. *Journal of Heredity* **78**, 134–42.

Wallace, B. (1989). One selectionist's perspective. *Quarterly Review of Biology* **64**, 127–45.

Walls, M. and Ketola, M. (1989). Effects of predator-induced spines on individual fitness in *Daphnia pulex*. *Limnology and Oceanography* **34**, 390–6.

Ward, P.J. (1989). A model of end-directedness. In *Issues in evolutionary epistemology* (ed. K. T. Hahlweg and C. A. Hooker), pp. 357–90, State University of New York Press, Albany.

Warner, R. R. (1975). The adaptive significance of sequential hermaphroditism in animals. *American Naturalist* **109**, 61–82.

Washburn, J. O., Gross, M. E., Mercer, T. R. and Anderson, J. R. (1988). Predator-induced trophic shift of a free-living ciliate: Parasitism of mosquito larvae by their prey. *Science* **240**, 1193–5.

Waters, C. K. (1991). Tempered realism about the force of selection. *Philosophy of Science* **58**, 553–73.

Watson, P.J. (1986). Transmission of a female sex pheromone thwarted by males in the spider *Linyphia litigosa* (Linyphiidae). *Science* **233**, 219–21.

Watt, W. B., Cassin, R. C. and Swan, M. S. (1983) Adaptation at specific loci. III. Field behavior and survivorship differences among *Colias* PGI genotypes are predictable from *in vitro* biochemistry. *Genetics* **103**, 725–39.

Webb, G. J. W. and Cooper-Preston, H. (1989). Effects of incubation temperature on crocodiles and the evolution of reptilian oviparity. *American Zoologist* **29**, 953–71.

Werner, E. E. (1986). Amphibian metamorphosis: growth rate, predation risk, and the optimal size at transformation. *American Naturalist* **128**, 319–41.

Werner, E. E. and Gilliam, J. F. (1984). The ontogenetic niche and species interactions in size-structured populations. *Annual Review of Ecology and Systematics* **15**, 393–425.

West-Eberhard, M. J. (1983). Sexual selection, social competition, and speciation. *Quarterly Review of Biology* **58**, 155–83.

West-Eberhard, M. J. (1984). Sexual selection, competitive communication, and species-specific signals in insects. In *Insect communication* (ed. T. Lewis), pp. 283–324, Academic Press, New York.

West-Eberhard, M. J. (1986). Alternative adaptations, speciation, and phylogeny (A review). *Proceedings of the National Academy of Sciences (U.S.A.)* **83**, 1388–92.

West-Eberhard, M. J. (1987) Flexible strategy and social evolution. In *Animal societies: theories and facts* (ed. Y. Ito, G. L. Brown and J. Kikkawa), pp. 35–51, Japan Scientific Societies Press, Tokyo.

West-Eberhard, M. J. (1989). Phenotypic plasticity and the origins of diversity. *Annual Review of Ecology and Systematics* **20**, 249–78.

Whiten, A. and Byrne, R. A. (1988). Taking (Machiavellian) intelligence apart: editorial. In *Machiavellian intelligence: social expertise and the evolution of intellect in monkeys, apes and humans* (ed. R. Byrne and A. Whiten), pp. 50–5, Oxford University Press.

Wiklund, C. (1990). Offspring protection by merlin *Falco columbaris* females: The importance of brood size and expected offspring survival for defense of young. *Behavioral Ecology and Sociobiology* **26**, 217–23.

Wilbur, H. M. and Fauth, J. E. (1990). Experimental aquatic food webs: Interactions between two predators and their prey. *American Naturalist* **135**, 176–204.

Wilkinson, M. (1990). A commentary on Ridley's cladistic solution to the species problem. *Biology and Philosophy* **5**, 433–46.

Williams, G. C. (1964). Measurement of consocation among fishes and comments on the evolution of schooling. *Michigan State University Museum Publications, Biology Series* **2**, 351–83.

Williams, G. C. (1966a). *Adaptation and natural selection*. Princeton University Press.

Williams, G. C. (1966b). Natural selection, the costs of reproduction, and a refinement of Lack's principle. *American Naturalist* **100**, 687–90.

Williams, G. C. (1979). The question of adaptive sex ratio in outcrossed vertebrates. *Proceedings of the Royal Society of London B* **205**, 567–80.

Williams, G. C. (1985). A defense of reductionism in evolutionary biology. *Oxford Surveys in Evolutionary Biology* **2**, 1–27.

Williams, G. C. (1989). A sociobiological expansion of evolution and ethics. In *Evolution and ethics* (ed. J. Paradis and G. C. Williams), pp. 179–214, 228–36, Princeton University Press, Princeton.

Williams, G. C. (1992). Mother Nature is a wicked old witch. *Quarterly Review of Biology* **67** (in press).

Williams, G. C. and Nesse, R. M. (1991). The dawn of Darwinian medicine. *Quarterly Review of Biology* **66**, 1–22.

Williams, G. C. and Taylor, P. D. (1987). Demographic consequences of natural selection. In *Evolution of Longevity in Animals*, 34th Brookhaven Symposium in Biology (ed. A. D. Woodhead and K. H. Thompson), pp. 235–45, Plenum, New York.

Wilson, D. S. (1980). *The natural selection of populations and communities*. Benjamin/Cummings Publishing Company, Menlo Park, CA.

Wilson, D. S. (1983). The group selection controversy: History and current status. *Annual Review of Ecology and Systematics* **14**, 159–87.

Wilson, D. S. (1986). Adaptive indirect effects. In *Community ecology* (ed. J.Diamond and T. J. Case), pp. 437–44, Harper and Row, New York.

Wilson, D. S. (1987). Altruism in Mendelian populations derived from sibling groups: The haystack model revisited. *Evolution* **41**, 1059–70.

Wilson, D. S. (1988). Holism and reductionism in evolutionary biology. *Oikos* **53**, 269–73.

Wilson, D. S. (1989). Diversification of single gene pools by density- and frequency-dependent selection. In *Speciation and its consequences* (ed. D. Otte and J A. Endler), pp. 366–85, Sinauer, Sunderland, MA.

Wilson, D. S. and Knollenberg, W. G. (1987). Adaptive indirect effects: the fitness of burying beetles with and without their phoretic mites. *Evolutionary Ecology* **1**, 139–59.

Wilson, D. S. and Sober, E. (1989) Reviving the superorganism. *Journal of Theoretical Biology* **136**, 337–56.

Wilson, E. O. (1975). *Sociobiology: The new synthesis*. Harvard University Press, Cambridge, MA.

Wimsatt, W. C. (1986). Developmental constraints, generative entrenchment, and the innate-acquired distinction. In *Integrating scientific distinctions* (ed. W. Bechtel), pp. 185–208. Martinus-Nijhoff, Dordrecht.

Wimsatt, W. C. and Schank, J. C. (1989). Two constraints on the evolution of complex adaptations and the means for their avoidance. In *Evolutionary progress* (ed. M. H. Nitecki), pp. 231–73, University of Chicago Press.

Wray, G. A. and McClay, D. R. (1989). Molecular heterochronies and heterotopies in early echinoid development. *Evolution* **43**, 803–13.

Wren, H. N., Johnson, J. L. and Cochran, D. G. (1989). Evolutionary inferences from a comparison of cockroach nuclear DNA and DNA from their fat-body and egg endosymbionts. *Evolution* **43**, 276–81.

Wright, J. and Cuthill, I. (1990). Manipulation of sex differences in parental care: the effect of brood size. *Animal Behavior* **40**, 462–71.

Wright, S. (1932). The roles of mutation, inbreeding, crossbreeding, and selection in evolution. *Proceedings of the VI International Congress of Genetics* **I**, 356–66.

Wynne-Edwards, V. C. (1962). *Animal dispersion in relation to social behavior.* Oliver and Boyd, Edinburgh.

Yodzis, P. (1981). The structure of assembled communities. *Journal of Theoretical Biology* **92**, 103–17.

Yoo, B. H. (1980). Long-term selection for a quantitative character in large replicate populations of *Drosophila melanogaster.* I. Response to selection. *Genetical Research,* **35**, 1–17.

Yoshimura, J. and Shields, W. M. (1987). Probabilistic optimization of phenotype distributions: a general solution for the effects of uncertainty on natural selection? *Evolutionary Ecology* **1**, 125–38.

Zeng, Z.-B. (1988). Long-term correlated response, interpopulation covariation, and interspecific allometry. *Evolution* **42**, 363–74.

Zinganov, V., Golovatjuk, G. J., Savvaitova, K. A. and Bugaev, V. F. (1987). Genetically isolate sympatric forms of threespine stickleback, *Gasterosteus aculeatus,* in Lake Azabachije (Kamchatka-Penninsula, USSR). *Environmental Biology of Fishes* **18**, 241–7.

Appendix: Excerpts from some classic works on adaptation

I believe that the following excerpts from works by Galen (May, 1968) and by William Paley (1836) are worth close attention by all biologists. Both illustrate how adaptation can be explored by seeking conformities to *a priori* design specifications. Galen's and Paley's main arguments are thoroughly mechanistic, and their occasional errors arise from the primitive state of physical science in their times, and from their total ignorance of natural selection and historicity. Thus Paley could show that the lens contributes to vision by focusing a sharp image on the retina, but he had no scientific understanding of what vision was designed to accomplish, and no way of knowing that some features of the eye might be nonadaptive historical legacies.

I hope that these excerpts will induce many readers to read the whole books. Many other historically significant works might also be read with profit by modern biologists. Galen and Paley built heavily on the works of Aristotle, now widely available in translation. A more recent classic that everyone ought to read is H. J. Muller's *Evidence of the Precision of Genetic Adaptation* (Muller, 1948). He used commonly studied mutations to establish that normal wild-type characters of *Drosophila melanogaster* are optimized. He did this by showing that the mutant forms, even those with seemingly trivial effects on the number and shape and distribution of bristles and wing veinlets, are functionally impaired by these departures from their optima. His main focus was an exploration of the elaborate machinery of dosage compensation whereby both females, with two X-chromosomes, and males, with one, normally develop almost exactly the same phenotypes for characters affected by genes on that chromosome.

Muller's essay is of 65 pages, far too long to include as part of this appendix, and I was unable to extract any small part that would capture the flavor of the whole tightly reasoned argument. It has a surprisingly modern ring after 44 years. While reading it I find it hard to bear in mind that he had no knowledge of modern molecular and developmental genetics, and no explicit understanding of Hamilton's (1964) concept of *inclusive fitness* or any of the recent work on life-history evolution. His discussion, like so many of Fisher's, Haldane's, and Wright's between 1930 and 1960, shows that theories are often understood and used effectively long before they are formalized. Despite the decades of progress on molecular and developmental genetics, Muller's work is still largely valid, even with respect to the technical details of dosage compensation in *Drosophila*.

Excerpts from Galen's Chapter 6 (*The hand*) of Volume 1 of 'The usefulness of the parts of the body' in May's (1968) translation

Come now, let us investigate this very important part of man's body, examining it to determine not simply whether it is useful or whether it is suitable for an intelligent animal, but whether it is in every respect so constituted that it would not have been better had it been made differently. One and indeed the chief characteristic of a prehensile instrument constructed in the best manner is the ability to grasp readily anything of whatever size or shape that man would naturally want to move. For this purpose, then, which was better— for the hand to be cleft into many divisions or to remain wholly undivided? Or does this need any discussion other than the statement that if the hand remained undivided, it would lay hold only on the things in contact with it that were of the same size as it happend to be itself, whereas, being subdivided into many members, it could easily grasp masses much larger than itself, and fasten accurately upon the smallest objects? For larger masses, the hand is extended, grasping them with the fingers spread apart, but the hand as a whole does not try to grasp the smallest objects, for they would escape if it did; the tips of two fingers are enough to use for them. Thus the hand is most excellently constituted for a firm grasp of things both larger and smaller than itself. Furthermore, if it was to be able to lay hold on objects of many different shapes, it was best for it to be divided into many differing members, as it now is, and for this purpose the hand is obviously adapted best of all prehensile instruments. Indeed, it can curve itself around a spherical body, laying hold of and encircling it from all sides; it surrounds firmly objects with straight or concave sides; and if this be true, then it will also clasp objects of all shapes, for they are all made up of three kinds of lines, convex, concave, and straight. Since, however, there are many bodies whose mass is too great for one hand alone to grasp, Nature made each the ally of the other so that both together, grasping such a body on opposite sides, are in no way inferior to one very large hand. For this reason, then, they face toward one another, since each was made for the sake of the other, and they have been formed equal to one another in every respect, a provision suitable for instruments which are to share the same action. Now when you have considered the largest objects that man can handle with both hands, such as a log or rock, then give heed, pray, to the smallest, such as a millet seed, a very slender thorn, or hair, and then, when you have considered besides how very many bodies there are that range in size from the largest to the smallest, think of all this and you will find that man handles them all as well as if his hands had been made for the sake of each one of them alone. He takes hold of very small objects with the tips of two fingers, the thumb and forefinger, and slightly larger objects with the same two fingers, but not with just the tips;

those still larger he grasps with three fingers, the thumb, forefinger, and middle finger, and if there are any larger yet, with four, and next, with five. After that the whole hand is used, and for still larger objects the other hand is brought up. The hand could act in none of these ways if it were not divided into fingers differently formed; for it was not enough in itself for the hand merely to be divided. What if there had been no finger opposing the four, as there is now, but all five of them had been produced side by side in one straight line? Is it not very clear that mere number could be useless, since an object to be held firmly must be either encircled from all sides or at least laid hold of from two opposite points? The ability to hold an object firmly would be destroyed if all the fingers had been produced side by side in one straight line, but as it is, with one finger set opposite the rest, this ability is nicely preserved; for this one finger has such a position and motion that by turning very slightly it acts with each of the four set opposite to it. Hence it was better for the hands to act as they do now, and Nature therefore gave them a stucture suited to such actions.

Now it was necessary not only that the tips of two opposed fingers should act in fastening upon small objects, but that the tips should also be such as they now are, soft, round, and provided with nails. Thus if the ends of the fingers were composed not of flesh but of bone, it would be impossible for them ever to lay hold of small articles such as thorns or hairs, nor would this be possible if, though they were fleshy, the flesh were too soft and moist. For it is necessary for the object grasped to be surrounded as completely as possible by the prehensor so that there may be firm support. No hard or bony substance can enfold anything, whereas substances moderately soft and therefore sufficiently yielding are able to do so. But of course excessively soft and almost fluid substances yield to hard objects more than is necessary and easily flow away from them. Consequently, the best instruments for grasping firmly would be those which, like the ends of the fingers, have a nature midway between hardness and extreme softness.

But since the objects themselves that are to be grasped also differ in consistency, some being harder or softer than others, Nature gave the finger tips a structure suited to laying hold of all kinds of them. This is the reason why the finger tips were not composed of either nail or flesh alone, but of both together, having the best mutual arrangement. Thus the flesh was located in the parts that face each other, by the extremities of which they were to seize on objects, and the nails were placed on the outside as a foundation for the flesh. The finger tips, then, lay hold on soft bodies with their fleshy parts alone, but are unable to pick up without the aid of the nails bodies that are hard and hence press and bear hard against the flesh; for then the flesh is turned back and needs a foundation. Furthermore, hard objects could not be grasped by the nails alone, because the nails, being hard, would readily slip off from them. Therefore, since the fleshy substance of the finger tips compensates for the slipperiness of the nails, and the nails offer support to the easily deformed flesh, the finger has been fashioned into an instrument capable of laying hold of all small, hard objects. You will understand more clearly what I mean when you consider nails that are out of proportion. Those that are excessively long and so strike against each other cannot pick up a little thorn or hair or anything else of the kind, and those that are too short to reach to the ends of the fingers deprive the flesh of its support and make it incapable of laying hold on anything. Only nails that come even with the ends of the fingers will best provide the service for the sake of which they were made. It is for this reason that Hippocrates[15] too has said, "The nails neither to project beyond, nor to fall short of, the finger tips." Thus, it is when they have a duly proportioned size that they best fulfill the uses for which they were made. Of course, they are also useful for many other purposes; for example, when it is necessary to scrape, scratch, skin, or tear something apart. In fact, in nearly all the circumstances of life and in all the arts, especially those requiring precise manual skill, we need some instrument of the sort, but it is as a prehensile instrument for seizing small, hard objects that the hand most needs the fingernails.

Excerpts from William Paley's Chapters 1 and 3 of Volume 1 of 'Natural Theology' (1836 edition)

IN crossing a heath, suppose I pitched my foot against a *stone*, and were asked how the stone came to be there: I might possibly answer, that for any thing I knew to the contrary, it had lain there for ever: nor would it perhaps be very easy to shew the absurdity of this answer. But suppose I had found a *watch* upon the ground, and it should be inquired how the watch happened to be in that place; I should hardly think of the answer which I had before given, that, for any thing I knew, the watch might have always been there. Yet why should not this answer serve for the watch as well as for the stone? why is it not as admissible in the second case, as in the first? For this reason, and for no other, viz, that, when we come to inspect the watch, we perceive (what we could not discover in the stone) that its several parts are framed and put together for a purpose, *e.g.* that they are so formed and adjusted as to produce motion, and that motion so regulated as to point out the hour of the day; that, if the different parts had been differently shaped from what they are, of a different size from what they are, or placed after any other manner, or in any other order, than that in which they are placed, either no motion at all would have been carried on in the machine, or none that would have answered the use that is now served by it. To reckon up a few of the plainest of these parts, and of their offices, all tending to one result:—We see a cylindrical box containing a coiled elastic spring, which, by its endeavour to relax itself, turns round the box. We next observe a flexible chain (artificially wrought for the sake of flexure), communicating the action of the spring from the box to the fusee. We then find a series of wheels, the teeth of which catch in, and apply to each other, conducting the motion from the fusee to the balance, and from the balance to the pointer; and at the same time by the size and shape of those wheels so regulating that motion, as to terminate in causing an index, by an equable and measured progression, to pass over a given space in a given time. We take notice that the wheels are made of brass in order to keep them from rust; the springs of steel, no other metal being so elastic; that over the face of the watch there is placed a glass, a material employed in no other part of the work, but in the room of which, if there had been any other than a transparent substance, the hour could not be seen without opening the case. This mechanism being observed (it requires indeed an examination of the instrument, and perhaps some previous knowledge of the subject, to perceive and understand it; but being once, as we have said, observed and understood), the inference, we think, is inevitable, that the watch must have had a maker; that there must have existed, at some time, and at some place or other, an artificer or artificers, who formed it for the purpose which we find it actually to answer; who comprehended its construction and designed its use.

I know no better method of introducing so large a subject, than that of comparing a single thing with a single thing; an eye, for example, with a telescope. As far as the examination of the instrument goes, there is precisely the same proof that the eye was made for vision, as there is that the telescope was made for assisting it. They are made upon the same principles; both being adjusted to the laws by which the transmission and refraction of rays of light are regulated. I speak not of the origin of the laws themselves; but such laws being fixed, the construction, in both cases, is adapted to them. For instance; these laws require, in order to produce the same effect, that the rays of light, in passing from water into the eye, should be refracted by a more convex surface, than when it passes out of air into the eye. Accordingly we find that the eye of a fish, in that part of it called the crystalline lens, is much rounder than the eye of terrestrial animals. What plainer manifestation of design can there be than this difference? What could a mathematical instrument maker have done more, to shew his knowledge of this principle, his application of that knowledge, his suiting of his means to his end; I will not say to display the compass or excellence of his skill and art, for in these all comparison is indecorous, but to testify counsel, choice, consideration, purpose?

To some it may appear a difference sufficient to destroy all similitude between the eye and the telescope, that the one is a perceiving organ, the other an unperceiving instrument. The fact is, that they are both instruments.— And, as to the mechanism, at least as to mechanism being employed, and even as to the kind of it, this circumstance varies not the analogy at all. For, observe what the constitution of the eye is. It is necessary, in order to produce distinct vision, tht an image or picture of the object be formed at the bottom of the eye. Whence this necessity arises, or how the picture is connected with the sensation, or contributes to it, it may be difficult, nay, we will confess, if you please, impossible for us to search out. But the present question is not concerned in the inquiry. It may be true, that, in this, and in other instances, we trace mechanical contrivance a certain way: and that then we come to something which is not mechanical, or which is inscrutable. But this affects not the certainty of our investigation, as far as we have gone. The difference between an animal and an automatic statue, consists in this,—that, in the animal, we trace the mechanism to a certain point, and then we are stopped; either the mechanism becoming too subtile for our discernment, or something else beside the known laws of mechanism taking place; whereas, in the automaton, for the compratively few motions of which it is capable, we trace the mechanism throughout. But, up to the limit, the reasoning is as clear and certain in the one case as in the other. In the example before us, it is a matter of certainty, because it is a matter which experience and observation demonstrate, that the formation of an image at the bottom of the eye is necessary to perfect vision. The image itself can be shewn. Whatever affects the distinctness of the image, affects the distinctness of the vision. The formation then of such an image being necessary (no matter how) to the sense of sight, and to the exercise of that sense, the apparatus by which it is formed is constructed and put together, not only with infinitely more art, but upon the self-same principles of art, as in the telescope or the camera obscura. The perception arising from the image may be laid out of the question; for the production of the image, these are instruments of the same kind. The end is the same; the means are the same.—The purpose in both is alike; the contrivance for accomplishing that purpose is in both alike. The lenses of the telescope, and the humours of the eye, bear a complete resemblance to one another, in their figure, their position, and in their power over the rays of light, viz. in bringing each pencil to a point at the right distance from the lens; namely; in the eye, at the exact place where the membrane is spread to receive it. How is it possible, under circumstances of such close affinity, and under the operation of equal evidence, to exclude contrivance from the one, yet to acknowledge the proof of contrivance having been employed, as the plainest and clearest of all propositions, in the other?

The resemblance between the two cases is still more accurate, and obtains in more points than we have yet represented, or that we are, on the first view of the subject, aware of. In dioptric telescopes, there is imperfection of this nature. Pencils of light, in passing through glass lenses, are separated into different colours, thereby tinging the object, especially the edge of it, as if it were viewed through a prism. To correct this inconvenience had been long a desideratum in the art. At last it came into the mind of a sagacious optician, to inquire how this matter was managed in the eye; in which there was exactly the same difficulty to contend with as in the telescope. His observation taught him, that, in the eye, the evil was cured by combining lenses composed of different substances, *i.e.* of substances which possess different refracting powers. Our artist borrowed thence his hint; and produced a correction of the defect, by imitating, in glasses made from different materials, the effects of the different humours through which the rays of light pass before they reach the bottom of the eye. Could this be in the eye without purpose, which suggested to the optician the only effectual means of attaining that purpose?

But farther; there are other points, not so much perhaps of strict resemblance between the two, as of superiority of the eye over the telescope; yet of a superiority which, being founded in the laws that regulate both, may furnish topics of fair and just comparison. Two things were wanted to the eye, which were not wanted (at least in the same degree) to the telescope; and these were, the adapttion of the organ, first, to different degrees of light; and, secondly, to the vast diversity of distance at which objects are viewed by the naked eye, viz. from a few inches to as many miles. These difficulties present not themselves to the maker of the telescope. He wants all the light he can get; and he never directs his instrument to objects near at hand. In the eye, both these cases were to be provided for; and for the purpose of providing for them a subtile and appropriate mechanism is introduced:

I. In order to exclude excess of light, when it is excessive, and to render objects visible under obscurer degrees of it, when no more can be had, the hole or aperture in the eye, through which the light enters, is so formed, as to contract or dilate itself for the purpose of admitting a greater or less number of rays at the same time. The chamber of the eye is a camera obscura, which, when the light is too small, can enlarge its opening; when too strong, can again contract it; and that without any other assistance than that of its own exquisite machinery. It is farther also, in the human subject, to be observed, that this hole in the eye which we call the pupil, under all its different dimensions, retains its exact circular shape. This is a structure extremely artificial. Let an artist only try to execute the same; he will find that his threads and strings must be disposed with greast consideration and contrivance, to make a circle, which shall continually change its diameter, yet preserve its form. This is done in the eye by an application of fibres, *i.e.* of strings, similar, in their position and action, to what an artist would and must employ, if he had the same piece of workmanship to perform.

II. The second difficulty which has been stated, was the suiting of the same organ to the perception of objects that lie near at hand, within a few inches, we will suppose, of the eye, and of objects which are placed at a considerable distance from it, that, for example, of as many furlongs (I speak in both cases of the distance at which distinct vision can be exercised). Now this, according to the principles of optics, that is, according to the laws by which the transmission of light is regulated (and these laws are fixed), could not be done without the organ itself undergoing an alteration, and receiving an adjustment, that might correspond with the exigency of the case, that is to say, with the different inclination to one another under which the rays of light reached it. Rays issuing from points placed at at a small distance from the eye, and which consequently must enter the eye in a spreading or diverging order, cannot, by the same optical instrument in the same state, be brought to a point, *i.e.* be made to form an image, in the same place with rays proceeding from objects situated at a much greater distance, and which rays arrive at the eye in directions nearly (and physically speaking) parallel. It requires a rounder lens to do it. The point of concourse behind the lens must fall critically upon the retina, or the vision is confused; yet, other things remaining the same, this point, by the immutable properties of light, is carried farther back when the rays proceed from a near object, than when they are sent from one that is remote. A person who was using an optical instrument, would manage this matter by changing, as the occasion required, his lens or his telescope; or by adjusting the distance of his glasses with his hand or his screw: but how is this to be managed in the eye? What the alteration was, or in what part of the eye it took place, or by what means it was effected (for if the known laws which govern the refraction of light be maintained, some alteration in the state of the organ there must be), had long formed a subject of inquiry and conjecture. The change, though sufficient for the purpose, is so minute as to elude ordinary observation. Some very late discoveries, deduced from a laborious and most accurate inspection of the structure and operation of the organ, seem at length to have ascertained the mechanical alteration which the parts of the eye undergo. It is found, that by the action of certain muscles, called the straight muscles, and which action is the most advantageous that could be imagined for the purpose,—it is found, I say, that whenever the eye is directed to a near object, three changes are produced in it at the same time, all severally contributing to the adjustment required. The cornea, or outermost coat of the eye, is rendered more round and prominent; the crystalline lens underneath is pushed forward; and the axis of vision, as the depth of the eye is called, is elongated. These changes in the eye vary its power over the rays of light in such a manner and degree as to produce exactly the effect which is wanted, viz. the formation of an image *upon the retina*, whether the rays come to the eye in a state of divergency, which is the case when the object is near to the eye, or come parallel to one another, which is the case when the object is placed at a distance. Can any thing be more decisive of contrivance than this is? The most secret laws of optics must have been known to the author of a structure endowed with such a capacity of change. It is as though an optician, when he had a nearer object to view, should *rectify* his instrument by putting in another glass, at the same time drawing out also his tube to a different length.

Observe a new-born child first lifting up its eyelids. What does the opening of the curtain discover? The anterior part of two pellucid globes, which, when they come to be examined, are found to be constructed upon strict optical principles; the self-same principles upon which we ourselves construct optical instruments. We find them perfect for the purpose of forming an image by refraction composed of parts executing different offices; one part having fulfilled its office upon the pencil of light, delivering it over to the action of another part; that to a third, and so onward; the progressive action depending for its success upon the nicest and minutest adjustment of the parts concerned; yet these parts so in fact adjusted, as to produce, not by a simple action or effect, but by a combination of actions and effects, the result which is ultimately wanted. And forasmuch as this organ would have to operate under different circumstances, with strong degrees of light, and with weak degrees, upon near objects, and upon remote ones; and these differences demanded, according to the laws by which the transmission of light is regulated, a corresponding diversity of structure; that the aperture, for example, through which the light passes, should be larger or less; the lenses rounder or flatter, or that their distance from the tablet, upon which the picture is delineated, should be shortened or lengthened: this, I say, being the case, and the difficulty to which the eye was to be adapted, we find its several parts capable of being occasionally changed, and a most artificial apparatus provided to produce that change. This is far beyond the common regulator of a watch, which requires the touch of a foreign hand to set it; but it is not altogether unlike Harrison's contrivance for making a watch regulate itself, by inserting within it a machinery, which, by the artful use of the different expansion of metals, preserves the equability of the motion under all the various temperatures of heat and cold in which the instrument may happen to be placed. The ingenuity of this last contrivance has been justly praised. Shall, therefore, a structure which differs from it, chiefly by surpassing it, be accounted no contrivance at all? or, if it be a contrivance, that it is without a contriver?

But this, though much, is not the whole: by different species of animals the faculty we are describing is possessed, in degrees suited to the different range of vision which their mode of life, and of procuring their food, requires. *Birds*, for instance, in general, procure their food by means of their beak; and, the distance between the eye and the point of the beak being small, it becomes necessary that they should have the power of seeing very near objects distinctly. On the other hand, from being often elevated much above the ground, living in the air, and moving through it with great velocity, they require, for their safety, as well as for assisting them in descrying their prey, a power of seeing at a great distance; a power of which, in birds of rapine, surprising examples are given. The fact accordingly is, that two peculiarities are found in the eyes of birds, both tending to *facilitate* the change upon which the adjustment of the eye to different distances depends. The one is a bony, yet, in most species, a flexible rim or hoop, surrounding the broadest part of the eye; which, confining the action of the muscles to that part, increases the effect of their lateral pressure upon the orb, by which pressure its axis is elongated for the purpose of looking at very near objects. The other is an additional muscle, called the marsupium, to draw, on occasion, the crystalline lens *back*, and to fit the same eye for the viewing of very distant objects. By these means, the eyes of birds can pass from one extreme to another of their scale of adjustment, with more ease and readiness than the eyes of other animals.

The eyes of *fishes* also, compared with those of terrestrial animals, exhibit certain distinctions of structure, adapted to their state and element. We have already observed upon the figure of the crystalline compensating by its roundness the density of the medium through which their light passes. To which we have to add, that the eyes of fish, in their natural and indolent state, appear to be adjusted to near objects, in this respect differing from the human eye, as well as those of quadrupeds and birds. The ordinary shape of the fish's eye being in a much higher degree convex than that of land animals, a corresponding difference attends its muscular conformation, viz. that it is throughout calculated for *flattening* the eye.

The *iris* also in the eyes of fish does not admit of contraction. This is a great difference, of which the probable reason is, that the diminished light in water is never too strong for the retina.

In the *eel*, which has to work its head through sand and gravel, the roughest and harshest substances, there is placed before the eye, and at some distance from it, a transparent, horny, convex case or covering, which, without constricting the sight, defends the organ. To such an animal, could any thing be more wanted, or more useful?

Thus, in comparing the eyes of different kinds of animals, we see, in their resemblances and distinctions, one general plan laid down, and that plan varied with the varying exigencies to which it is to be applied.

There is one property however, common, I believe, to all eyes, at least to all which have been examined, namely, that the optic nerve enters the bottom of the eye, not in the centre or middle, but a little on one side; not in the point where the axis of the eye meets the retina, but between that point and the nose. The difference which this makes is, that no part of an object is unperceived by both eyes at the same time.

Index

Adaptation
  compass of 39–40, 43–4
  demonstration of 40–1, 98, 101–5,
    112–13
Adaptationist program 5–6, 64–5
Adaptive performance 39, 104
Alarm call 115
Alarm substance 113–16
Albatross 139
Alexandria, Library at 10, 76
Allee, W. C. 47
Allometry 81–3
Allopatry 28, 121–2
Angiosperm ascendancy 34–5 (*see also*
  Plant)
Anglerfish 77–8
Antibiotic resistance 116
Antifreeze 138–9
Antlers 82, 122–3
Ascidian 124
Asexual reproduction 21–2
  clade selection against 31–2, 35–6, 148–9
Asymmetry, selection for 86–7
Australopithecus 122
Avatar 52, 120

Bacteria, streptomycin-resistant 116
Baptista's burden 149–50
Bats 78–9
Bauplan 87–8
Bdelloidea 148–9
Bear 79
Beaver 65
Beech 130
Bees 44, 48, 93–4
Beetle 85
Beryx 131
Bird
  alarm cells 115
  brood parasitism 20–1
  clutch size 66, 70
  communication 110, 112
  evolutionary rates 114, 128
  fitness 118
  flight 39, 41, 140
  flocking 45

foraging 65, 147
habitats 101–2
kin groups 20
parenting 65
Passeriform 35, 45, 109
polyandry 111
senescence 151–2
size 82, 84
song 109, 111–12
temperature 136–40
two-hearted 81
viviparity vs. oviparity 139–41
Black holes 36, 75, 79
Blood color 62, 64
Blue goose 123
Blue whale 76
Bluegill sunfish 40, 92, 94–7
Bovid 52
Brachiopod 26, 35, 125
Brachyraphis 96
Butterfly 57, 87, 123
  effect 6

Cambrian 31, 126
Camouflage 104
Caterpillar 91, 123
Catfish 40, 41, 114
Cause and effect 8–9
Cave fish 114
Cervid 52, 70, 82
Chaenichthyidae 88
Character correlations 64, 101–4
Character stasis 135–42
Chickadee 67
Chloroplast 42
Choking on food 7, 63
Christmas tree 11, 13, 15
Cichlid 90, 120, 123
Clade selection 23–37, 46–8, 120, 148–9
  compared to selection of genes 27–30
  importance 31–7
  normalizing 132–5
  on phytophagy 32–3
  on size 32–5, 50–3
  requirements 24–7
Cladocera 43, 94, 97

Cladogenesis 98–100 (*see also* Species; Speciation)
Clausen 46
Cliff edge effect 66–8
Cockroach 54, 107
Codex and codical domain 10–13, 23
Coefficient of relationship 19–21, 42–3
Commensals 54
Communication 110–13
Community 55, 130
Compass of adaptation 39–40
Computer 127
Constraint
  developmental 80–5
  genetic 64, 85–7
  phylogenetic 76–80, 135–42
Copepod 128
Coral 54–5
Cost of meiosis 148
Countergradient selection 98
Cretaceous 27, 131–2
Cricket 107
Crocodile 115, 140
Crustacea 76–7, 120, 125
Cultural evolution 15–16, 18
Cypriniformes 114

Danio 114
Daphnia 43, 94
Darwin and Darwinism 5, 8, 38, 86
  conceptual difficulties 106, 143
  on gradualism 76–81, 124–7
  variation in nature 56, 97–8
Deer 70, 82
Demerec, M. 116
Dendrogram 13–15, 18, 27–8, 34, 46, 53, 122, 133
Density effects 57, 146–7
Descartes 4
Design specifications 40–1, 43–4, 98, 101–5, 112–13
Disease resistance 27
DNA 11–12, 13–14, 23, 42 (*see also* Gene)
Dobzhansky, Th. 47, 99
Dogs, genotype discrimination by 12
Domains of selection 10–11, 13–14, 23
Domestic animals and plants 53, 85, 86, 130
Don Quixote 11, 13
Dretske's principle 13, 30
Driesch's entelechy 3–4
Driving chromosomes 41
Drosophila 40, 86, 92, 99, 107, 122–3, 125

Ear, of mammal 76

Echinoderm 85
Ecological species 120–1
Ecological succession 53
Ecotype 46
Eel 79
Electrolyte concentration 138–9
Embryology, and vitalism 3–4
English sparrow 129
Environmental abnormality 118
Epigenetic load 93
ESS 42, 44, 56–8, 63, 66, 72–6, 93–8, 136
Ethics 9
Evolution
  misunderstandings of 3, 7, 127–8
  rate of 127–8, 132–5
Evolutionarily significant unit 123
Evolutionarily stable strategy, *see* ESS
Exaptation 77
Extended phenotype 45
Extraterrestrial visitor 4–5
Eye (vertebrate) 72–4, 152–3, 196–201

Fallacies
  female pheromone 111–13
  species 120–5
Female choice 68, 106–10
Fiddler crab 86
Finch 97
Fish
  ancestors 77
  asexual 21
  cartilaginous 138
  cave dwellers 114–15
  communication 114
  electrolyte concentration 138–9
  evolutionary rates 128–9, 131
  fecundity 69–70
  fitness 93
  life history stages 85, 92
  mouth parts 90
  parenting 65
  polymorphism 84, 87, 92, 94–7, 120, 123
  reproduction 141
  schools 60, 115
  skeleton 131, 132, 135
  species distinctions 119
  vertebral count 79–80, 142
Fisher, R. A. 15, 17, 47, 107, 153
Fitness 16–17, 48, 61–71, 93, 107, 109, 118
Flour beetles 107
Flowering time 63
Foraminifera 32
Fossil record 32–5, 52–4, 64, 83, 86–7, 124–6, 132, 135
Frequency-dependent selection 44, 46, 56–60, 92

Fright reaction, 114–15
Frog 92, 132, 149

Galen 40–1, 189–94
Game theory, *see* ESS
Gastrotrich 149
Gene
  as level of selection 16–22, 27–9
  definition 11, 18
  frequency 18, 30
  in development 18
  object vs. information 11
  pool 16–17, 19, 23–7
  selfishness 15
Genealogy 19
General adaptation 34
Generative entrenchment 84–5
Genet 43, 44, 94
Genetic drift 29
Genetic environment 18
Genetic load 93, 143–8
Genotype as level of selection 16–17, 22
Geological time 42, 87, 127–8, 135
Gill arches 77
Giraffe 76
Goldfish 80–1
Goldschmidt, R. 128
Gonorhynchiformes 114
Gorilla 238
Gould, S. J. 3, 78
Gradualism 76–81, 124–7
Graptolite 43
Grazing 117–18
Great American Interchange 6, 8
Griffin, D. R. 4
Group-size optimization 60
Gull 124
Guppy 128, 129
Gymnotid 114

Hagfish, 138–9
Haldane, J. B. S. 47, 7
Haldane's dilemma 143–8
Hamilton–Zuk model 108
Hamlet's soliloquy 127
Hand, *see* Human hand
Harvey, W. 64
Haystack model 26, 48–9
Hearts, number of 81
Helpful stress effect 116–18
Hemoglobin 88
Hermaphrodite 149
Historicity 3, 5–9, 31, 34, 36–7, 71–8
Holism 9, 48–9
Hominidae 7, 29, 31, 39, 103, 121–2, 132,
  135

Homology 13
Human (*see also* Hominidae)
  adaptations 39, 144–5
  brain 78–9, 132, 152
  chin 78
  evolution 132, 144–5
  eye 152–3, 196–200
  fitness 94, 144–6
  hand 40, 60–1, 76, 192–4
  kidney 81
  obesity 118
  pigmentation 92–3
  size 103
  twinning 79
  variation 94
Huxley, J. S. 99
Hymenoptera 27, 84 (*see also* Insect,
  social)

Ilyodon furcidens, 120, 123
Inbreeding 21–2
Inclusive fitness 19
Indigo buntings 107
Insect
  evolutionary innovation 125
  diet and diversity 32–3
  glands 113
  larvae 45
  sex determination 150
  social 47–9, 54, 91, 93–4
  stages 92
Interactor 10, 12–13, 16, 23, 38–55
Ipomopsis 117–18
Irish elk 78
Irreversible evolution 36, 75, 79
Isthmus of Pañama 6, 8

Jaw 77–8
Johannsen 85
Jurassic 139

Key innovation 35
Killifish 21, 110
Kin recognition 20–1
Kin selection 19–21, 43, 45, 115

Lag load 136
Lamarkism 128
Lamprey 124
Larynx 63
Leech 142
Lek paradox 106–11
Lemming 49, 70, 137

Leopard frog 120, 124
Levels of selection 5–6, 16–18, 38, 43–7
   (*see also* Clade selection)
Library at Alexandria 10, 76
Lichen 54–5
Lilies 112
Lion 60
Lizard 52, 79, 112, 135, 141
Lobster 94

Machiavelli 15
Macroevolution 31, 23–37, 50–5, 125–42,
   132–5 (*see also* Clade selection)
Maize 143, 164
Malaria resistance 145
Male pheromone 113
Malthusian parameter 17, 146
Mammal
  African 26
  alarm calls, 115
  antlers 82, 122–3
  cervical vertebrae 73, 79, 141–2
  clade selection 32–4, 52–3
  coat color 103
  ears 76–7
  energy usage 68
  eyes 62
  gametes 74–5
  genome 145
  herbivory 117
  neck 7, 76, 79, 141–2
  parenting 65
  senescence 151
  size 94, 50–2, 70
  temperature requirements 74, 136–8
  testicle position 74–5, 137–8
Manipulation by parasite 117
Material domain 10, 23 (*see also* Codex
   and codical domain; Interactor)
Mayr, E. 47, 99, 118–19, 123–5
Mechanism (vs. vitalism) 3–5
Meiotic drive 41–2
Meme 10–11, 13–16
Mendel 7, 18
Meristic characters 73–80, 141–2
Mesozoic 125
Mimicry 57, 92, 95
Mind and mentalism 4–5
Minnow 45, 113–16
Mitochondria 42
Molecular evolution 34
Molidae 80
Mollusk 74, 85, 120–1 (*see also* Mussel;
   Pelecypods; Snail)
Monkey 127
Monocotyledon 81

Mosquitofish 128–9
Moth 28, 111–13
Mouse 137
Muller, H. J. 40–1, 65, 75, 153, 190–1
Muller's ratchet 75, 148–9
Mussel 44
Mutation 24, 27, 29–30, 86, 107
Mutualism 54–5

Narhwal tusk 86–7
Natural selection 5–6, 127 (*see also* Kin,
   and Clade selection)
  countergradient 98
  density-dependent 57
  for asymmetry 86–7
  frequency-dependent 56–60
  response to 97–8
Necks (of mammals) 7, 76, 79, 141–2
Nemoria 91
Neoplasm 43
Nervous system 72
Normalizing clade selection 132–5

Oaks 28
Ophyrotrocha 92
Optimal foraging 39, 45, 65
Optimization 15, 39, 56–7, 60–5, 92, 136
Ordovician 32
Organelle 41–2
Organism
  as document 6, 72–6
  object vs. pattern 11–12, 17–18
Osmoconformity 138
Ostariophysi 114–16
Outlaw genes 41

Paley 40, 72, 190, 195–201
Pan 29, 103, 122
Paradise fish 82
Parameter optimization 60–5
Parasite 54–5, 93–6, 117
Parasitoid 57, 75
Pasteur 113
Peacock 78
Pelecypods 26, 35, 44, 85, 125
Peripheral isolate 125–6
Phanerozoic 32
Pheasant 109
Phenotype 23, 45
Pheromones 106–11, 113
Phylad and phylad selection 23–37
Pickerel 45
Piranha 114–15
Plant (*see also* Pollen dispersal; Seed;
   Seedling competition)

crop 144
evolutionary innovation 125
grazing 117–18
kin selection 20
outcrossing 36, 135
population fitness 93
populations 20–1, 128
size 35, 69–70, 93
Platyfish 84
Play 68–9, 78
Pleiotropy 64
Pleistocene 53, 126, 129–32
Pliocene 8, 103, 131
Ploidy 40
Pollen dispersal 66–7, 70
Pollinators 64
Polymorphism 85, 90–1, 94–7, 120, 123–4
Polyplectron 109
Population
    adaptive regulation of 47–8
    as level of selection 46–8
    defined 23–4
    size 28–9
Poultry 82, 85
Primates 76–9, 87 (*see also* Gorilla;
    Hominidae; Human; Protohominid;
    Pan)
Prisoners' dilemma 59–60
Progress 34–5
Protist ancestor of mammals 77
Protochordate ancestors 63
Protohominid 78, 122
Protozoa 97
Punctuated equilibrium 14, 53, 125–6, 134

Quaternary 55

Rabbit 104
Racemose phylogeny 133–5
Ramet 43–4, 70, 94, 104
Recent 126, 131, 132
Recipe 18–19
Reductionism 9, 48–9
Redwinged blackbird 112
Replicator 10, 12, 16
Reproductive success 16–17, 39, 46, 48,
    94–5, 98, 149
Reptile 47, 88, 112, 139, 140–1, 150
Rhinogradentia 78
Rivulus 21
RNA 11–12
Rodent 50–2, 55
Root symbiont 55
Rotifer 148–9

Salamander 88, 112, 124–5, 135

Salmonid 25, 92, 95
Schreckstoff 113–16
Science fiction 7
Scorpionfly 41
Sculpin 80
Scurvy 90
Sea cow 76
Sea turtle 139, 141
Sea urchin 85
Seahorse 81
Seal 139
Seaweed fly 107
Seed 64, 69
Seedling competition 146–7
Segregation distorter 41–2
Senescence 150–2
Sex chromosome 75
Sex determination 149–50
Sex ratio 27, 56–7, 82, 149–50
Sexual selection 36, 44, 68, 78, 106–11,
    148
Sexy son model 107
Shark 138
Siluriformes 114
Silverside 149–50
Simpson, G. G. 47, 53
Siphonophore 43
Size
    animals generally 69–70, 93
    beans 85
    birds 82, 84
    hominids 103
    mammals 94, 50–3, 70
    plants 35, 69–70, 93
    selection for 32, 50–3, 69, 93, 109
Snail 27
Snake 73, 88, 112, 141
Snow goose 97–8, 123
Soft selection 146–7
Soil arthropod 25
Soma 17, 23
Spandrel 77–9
Speciation 98–100, 125–6
Species
    concept 118–20, 125
    ecological 120–1
    fallacies 118–26
    individuality 119–23
    selection 24, 53, 125–6
    trophic 120–1
Spider 113
Spruce 130
Stasis 126–43
Stebbins, G. L. 99
Stickleback 110, 112, 124, 132, 132–5
Stone Age 39, 62
Strawberry 70

Sturtevant, A. H. 47
Supernormal stimulus 108
Superorganism 48

Tadpole 92
Tails, of vertebrates 80–1
Tamarin 113
Tapeworm 45
Termite 54, 84, 112
Testicle position 74–5
Titanothere 52
Towhee 123
Trait group 19–20, 48–9
Transcription 12
Transposable element 41, 143
Tree 65, 70, 81, 123, 127
Trilobite 26, 35
Trophic unit 121
Tundra 130
Turing test 4
Turtle 139–41
Twinning 79

Ungulate 52
Unionid 85

Unity of type 87
Universal Darwinism 8
Urea 138

Variation, causes of 91–3
Vertebrate eye 72–4, 196–201
Vitalism 3–5
Viviparity 139–41
Von Baer's law 84–5
Von Frisch, K. 113

Warbler 128
Warburton, F. E. 127
Weasel 50–1
Widow bird 110
Winnings 69–70, 86, 93
Wolf 60
Wrasses 35

Xerox principle 13

Zebra danio 114
Zooxanthellae 54–5